Nature-Based Solutions in Supporting Sustainable Development Goals

Theory and Practice

Nature-Based Solutions in Supporting Sustainable Development Goals

Theory and Practice

Edited by

Haozhi Pan
Shanghai Jiao Tong University, Shanghai, China

Zahra Kalantari
Professor, KTH Royal Institute of Technology, Stockholm, Sweden

Carla Ferreira
Polytechnic Institute of Coimbra, Applied Research Institute, Coimbra, Portugal

Cong Cong
Urban Science and Planning, Massachusetts Institute of Technology, Cambridge, MA, United States

ELSEVIER

Elsevier
Radarweg 29, PO Box 211, 1000 AE Amsterdam, Netherlands
125 London Wall, London EC2Y 5AS, United Kingdom
50 Hampshire Street, 5th Floor, Cambridge, MA 02139, United States

Notices
Knowledge and best practice in this field are constantly changing. As new research and experience broaden our understanding, changes in research methods, professional practices, or medical treatment may become necessary.

Practitioners and researchers must always rely on their own experience and knowledge in evaluating and using any information, methods, compounds, or experiments described herein. In using such information or methods they should be mindful of their own safety and the safety of others, including parties for whom they have a professional responsibility.

To the fullest extent of the law, neither the Publisher nor the authors, contributors, or editors, assume any liability for any injury and/or damage to persons or property as a matter of products liability, negligence or otherwise, or from any use or operation of any methods, products, instructions, or ideas contained in the material herein.

ISBN: 978-0-443-21782-1

For information on all Elsevier publications
visit our website at https://www.elsevier.com/books-and-journals

Publisher: Jonathan Simpson
Acquisitions Editor: Jessica Mack
Editorial Project Manager: Ellie Barnett
Production Project Manager: Sruthi Satheesh
Cover Designer: Christian Bilbow

Typeset by STRAIVE, India

Working together
to grow libraries in
developing countries

www.elsevier.com • www.bookaid.org

Contents

Section 1
NbS to support sustainable development goals

1.1 Forests for climate change mitigation: Temporal dynamics of carbon sequestration in the forests of Stockholm County

Jessica Page, Victoria Kareflod, and Elisie Kåresdotter

Section 2
Policy approach to promote NbS

2.2 NbS and policy communication

Cong Cong and Diego Temkin

2.3 The integration and adoption of the concept of urban resilience into policy in the Netherlands

Nina Escriva Fernandez and Haozhi Pan

Contributors

Numbers in parenthesis indicate the pages on which the authors' contributions begin.

Julia J. Aguilera Rodriguez (129), Institute for Environmental Sciences, University of Geneva, Geneva, Switzerland

Nejc Bezak (25), Faculty of Civil and Geodetic Engineering, University of Ljubljana, Ljubljana, Slovenia

Luigia Brandimarte (53), Department of Sustainable Development, Environmental Science and Engineering, Sustainability Assessment and Management, KTH Royal Institute of Technology, Stockholm, Sweden

Cong Cong (153), Urban Science and Planning, Massachusetts Institute of Technology, Cambridge, MA, United States

Yashar Dadrasajirlou (83), Civil Engineering Department, Semnan University, Semnan, Iran

Elin Ekström (53), Department of Sustainable Development, Environmental Science and Engineering, Sustainability Assessment and Management, KTH Royal Institute of Technology, Stockholm, Sweden

Nina Escriva Fernandez (111,173), School of International and Public Affairs, Shanghai Jiao Tong University, Shanghai, China

Carla Sofia Santos Ferreira (53,83), Polytechnic Institute of Coimbra, Applied Research Institute; Research Centre for Natural Resources Environment and Society (CERNAS), Polytechnic Institute of Coimbra, Coimbra, Portugal

Alberto Fresolone (129), Equity and Justice Group, International Institute for Applied Systems Analysis, Laxenburg, Austria

Jiří Jakubínský (25), Global Change Research Institute CAS, Brno, Czech Republic

Zahra Kalantari (83), KTH Royal Institute of Technology, Stockholm, Sweden

Hojat Karami (83), Civil Engineering Department, Semnan University, Semnan, Iran

Victoria Kareflod (3), Department of Sustainable Development, Environmental Science and Engineering (SEED), School of Architecture and the Built Environment (ABE), KTH Royal Institute of Technology, Stockholm, Sweden

Elisie Kåresdotter (3), Department of Sustainable Development, Environmental Science and Engineering (SEED), School of Architecture and the Built Environment (ABE), KTH Royal Institute of Technology, Stockholm, Sweden

JoAnne Linnerooth-Bayer (129), Equity and Justice Group, International Institute for Applied Systems Analysis, Laxenburg, Austria

Juliette C.G. Martin (129), Equity and Justice Group, International Institute for Applied Systems Analysis, Laxenburg, Austria

Seyedali Mirjalili (83), Artificial Intelligence Research and Optimization Centre, Torrens University, Adelaide, SA, Australia

Jessica Page (3), Department of Physical Geography, Stockholm University, Stockholm, Sweden

Haozhi Pan (111,173), Shanghai Jiao Tong University, Shanghai, China

Alireza Rezaei (83), Civil Engineering Department, Semnan University, Semnan, Iran

Anna Scolobig (129), Institute for Environmental Sciences, University of Geneva, Geneva, Switzerland

Mojca Šraj (25), Faculty of Civil and Geodetic Engineering, University of Ljubljana, Ljubljana, Slovenia

Diego Temkin (153), Department of Urban Studies and Planning, MIT, Cambridge, MA, United States

Kristina Unger (25), Faculty of Civil and Geodetic Engineering, University of Ljubljana, Ljubljana, Slovenia

About the editors

Haozhi Pan, Professor, Shanghai Jiao Tong University, Shanghai, China

Haozhi Pan integrates regional economics and spatial data science to explore how cities can achieve economic growth and digital smartness. His work also focuses on co-governance goals such as carbon neutrality and citizen participation.

Zahra Kalantari, Professor, KTH Royal Institute of Technology, Stockholm, Sweden

Zahra Kalantari has successfully led and carried out interdisciplinary research with a focus on understanding of earth and human systems to develop science, technology, and innovation solutions to the planet's most pressing environmental challenges associated with the combined effects of changes in climate, land use, and water use in terrestrial environments.

Carla Ferreira, Auxiliary Researcher, Polytechnic Institute of Coimbra, Applied Research Institute, Coimbra, Portugal

Carla Ferreira's main research interests include land degradation and restoration, impacts of anthropogenic activities and climate change on land at different scales, nature-based solutions and ecosystem services, and sustainable land-use planning.

Cong Cong, Lecturer, Urban Science and Planning, Massachusetts Institute of Technology, Cambridge, MA, United States

Cong Cong's research focuses on using data, computing, and informatic tools to inform more effective decision-making. She explores the intersection of computer science and urban planning and uses large-scale urban modeling and spatial big data analysis as exploratory techniques.

Introduction: Nature-based solutions in supporting sustainable development goals

Haozhi Pan[a], Carla Ferreira[b], Zahra Kalantari[c], and Cong Cong[d]

[a]Shanghai Jiao Tong University, Shanghai, China, [b]Polytechnic Institute of Coimbra, Applied Research Institute, Coimbra, Portugal, [c]KTH Royal Institute of Technology, Stockholm, Sweden, [d]Urban Science and Planning, Massachusetts Institute of Technology, Cambridge, MA, United States

1 Introduction

The United Nations launched the sustainable development goals (SDGs) in 2015 as part of the 2030 Agenda for Sustainable Development, marking a shift from the previous Millennium Development Goals (United Nations General Assembly, 2015). Central to achieving these goals are nature-based solutions (NbS), which emphasize the sustainable management of natural resources and ecosystems to support economic activities and enhance community well-being. NbS encompass a diverse array of conservation and sustainability measures that efficiently address societal challenges, delivering economic, social, and environmental benefits (Martín et al., 2020). For instance, urban NbS such as green spaces provide substantial health advantages, including the reduction of cardiovascular diseases and obesity, thereby contributing to SDG 3.4 (James et al., 2015; Tate et al., 2024).

The implementation of NbS in complex social-ecological systems involves intricate dynamics such as nonlinearities, feedback loops, and evolving interconnections (Zhang et al., 2021). Implementations play a crucial role in navigating trade-offs and creating synergies among the co-benefits associated with multiple SDGs. For example, initiatives like establishing green alleys and planting trees not only manage storm water effectively but also provide additional benefits such as groundwater recharge, reduction of urban heat islands, and creation of habitats for wildlife (Newell et al., 2013). Furthermore, NbS strategies can evolve over time due to natural processes or external pressures like climate change, highlighting the need for an intertemporal approach to managing these trade-offs and synergies (Martín et al., 2020). Understanding and responding to these dynamics are essential for balancing social, economic, and environmental objectives, thereby enhancing the overall effectiveness of NbS in contributing to the SDGs.

This book delves into the intricate connections between the NbS and the SDGs. While it cannot comprehensively cover every interaction among these diverse goals, our focus is on highlighting key areas, particularly those relevant to climate and agriculture. Through presenting evidence-based discussions and analysis, we examine the nuanced links between NbS and the SDGs. Moreover, our ambition extends beyond mere identification of connections; we aim to propose feasible policy actions. By offering pragmatic insights into implementing, communicating, and scaling up the linkages between NbS and SDGs, this book aims to strengthen their relationship. It provides grounded evidence and actionable policy instruments to foster meaningful contributions toward achieving the SDGs through NbS.

This book is structured in two main sections. The first section presents a state of the art about the impact of different NbS in supporting several SDGs and includes case studies in distinct environments, including NbS applications in forest and urban areas, as well as mixed land use catchments. It discusses pathways and mechanisms for NbS to reduce GHG emissions and adapt to climate change, assesses the impact of real-life NbS projects, and presents pathways to enabling successful implementation and upscaling of NbS. This section includes five chapters. The second section of the book focuses on policy aspects, discussing current governance models and co-creation approaches to enhance social benefits while providing ecological and economic benefits. This section comprises three chapters.

2 NbS and SDG 2 (Zero Hunger), SDG 15 (Life on Land)

Achieving the Zero Hunger (SDG 2) target is still a global challenge. Projections indicate that 600 million people will still suffer from hunger by 2030 (Sachs et al., 2024). Land use systems are fundamental to achieving SDG 2, but over the last years they have been affected by external factors such as climate change and intensive anthropogenic activities that constrain their ability to feed a growing world population, deliver healthy diets, and sustainably maintain the environment. Agriculture occupies more than half of the Earth's land surface (Shukla et al., 2019) and uses 70% of freshwater resources (UNESCO, 2021). Climate change and the associated increase in frequency and intensity of extreme events (e.g., droughts and floods) and increasing water scarcity (IPCC, 2023), particularly in arid and semiarid regions (Ferreira et al., 2022), are relevant tests for the agriculture sector. At least 26% of the costs of damage from climate-related disasters are absorbed by the agricultural sector (FAO, 2015). Land degradation and desertification (i.e., land degradation in arid, semiarid, and dry subhumid areas) represent additional challenges to achieve SDG 2. Land degradation affects 40% of the planet and 3.2 billion people (Pereira et al., 2024) and is projected to affect 95% of the Earth's terrestrial area in the next 30 years (Sonneveld et al., 2018). Agricultural land is also a key aspect affecting SDG 15 (Life on Land). The conversion of natural areas into agriculture as well as afforested areas affects the land and biodiversity targets. These targets are based on the Kunming-Montreal Global Biodiversity Framework (2022), which aims to, e.g., halt the loss of land vital to biodiversity by 2030, implement biodiversity-friendly agricultural practices by achieving 50% of cropland under agroecological practices, and reach the zero loss of land where natural processes predominate (e.g., forests). Agriculture practices also affect SDG 15, aiming to *"protect, restore and promote sustainable use of terrestrial ecosystems, sustainably manage forests, combat desertification, and halt and reverse land degradation and halt biodiversity loss"* (Sachs et al., 2024).

NbS can support the achievement of both SDG 2 and SDG 15 by considering both natural functions and human applications to overcome challenges related to the functioning of the biosphere and support a transition to sustainable climate-resilient food systems (Kalantari et al., 2023). Keestra et al. (2023) propose three types of NbS to achieve sustainable development in food systems: (i) intrinsic NbS, those associated with better use of existing ecosystems; (ii) hybrid NbS that make use of adapted ecosystems; and (iii) inspired NbS that consist of newly designed and constructed ecosystems. Examples of intrinsic NbS include rainwater harvesting or runoff derived from a catchment area storage for irrigation, which supports, e.g., increasing agricultural productivity in areas with water shortage but also reduces flood risks by reducing runoff and mitigates erosion (Ferreira et al., 2023; Potocki et al., 2023). Examples of hybrid NbS include the combat of crop pests and diseases without the use of agrochemicals (e.g., intercropping, crop rotations, and creating natural refuges for predators that feed on pests), which provides multiple benefits such as healthier food and supports biodiversity decline (Maes & Jacobs, 2017). Inspired NbS comprise, for example, the reuse of treated wastewater for irrigation to reduce dependence from groundwater and surface water sources (Gonzalez-Flo et al., 2023). The implementation of successful NbS requires an improved understanding of the underlying mechanisms of NbS to support agriculture transformation (Keestra et al., 2023). Apart from agricultural land, numerous examples of NbS have been used for ecosystem restoration (e.g., constructed wetlands) and reverse degradation of ecosystems (e.g., reforestation), relevant to support achieving SDG 15. Demonstrating how NbS can support climate-resilient agriculture systems is of utmost importance to increase learning and support policies and governance mechanisms for NbS adoption.

The chapter from Ekström et al. (this book) discusses the role of wetlands in water management within catchments dominated by agricultural land use. Based on hydrological modeling, the authors investigate the impact of a large number of constructed wetlands on flood regulation in a large catchment located in southwest Sweden. Although the study demonstrates the relatively limited capacity of wetlands to support flood regulation, their water storage capacity may provide a relevant contribution to water availability during dry spells. Enhancing water availability to agricultural lands is particularly relevant in water-scarce regions, given the increasing conflicts between water uses and the typical priorities attributed to domestic water uses over agriculture irrigation.

3   NbS and SDG 11 (Sustainable Cities and Communities), SDG 13 (Climate Action)

Climate change mitigation and adaptation in urban areas represents a focal point in NbS research, prominently featured in our book. The European Union's commitment to achieving climate neutrality by 2030 and reducing net emissions underscores NbS's potential to support both SDG 11 and SDG 13. EU climate policies emphasize integrating NbS into urban planning to tackle interconnected challenges related to climate, biodiversity, and societal issues (European Commission, 2022). Incorporating NbS into policy initiatives underscores their role in advancing sustainable urban environments and enhancing climate resilience (Seddon et al., 2020). By embracing NbS, cities can emerge as centers of innovation and experimentation, showcasing the efficacy of these solutions in advancing SDGs and addressing the intricate relationship between urbanization and climate change (Pan et al., 2023).

NbS play a crucial role in both carbon mitigation and climate resilience enhancement. While traditional carbon sequestration methods focus on storing carbon in vegetation, soil, and wetlands, NbS offer broader benefits. Ecosystem services and green infrastructure strategies, such as urban agriculture and the greening of streetscapes, promote environmentally friendly behaviors and reduce reliance on fossil fuels. These initiatives encourage walking and cycling, curb urban sprawl, and reduce energy consumption for heating and cooling by regulating local climates (Pan et al., 2023). Concurrently, these behavioral changes support SDG 12 (Responsible Consumption and Production) by fostering pro-environmental lifestyles in tandem with NbS implementation. Additionally, NbS bolster urban resilience and enhance the quality of life. Measures like green roofs, rain gardens, and urban agriculture mitigate the impacts of extreme weather events by enhancing water retention, reducing stormwater runoff, and lowering urban temperatures through shading and evapotranspiration. These solutions also provide recreational spaces, esthetic value, and opportunities for social cohesion, thereby contributing to the overall well-being of urban residents. By integrating NbS, cities can strengthen their ability to adapt to climate change while fostering inclusive and sustainable urban development (Cortinovis et al., 2022).

Two chapters in this book discuss urban resilience and adaptation to climate change. Existing studies explore innovative methods for flood control in urban areas (SDG 11 and SDG 13), particularly those based on NbS. Among various urban flood management methods, green infrastructure, best management practices, and low-impact development (LID) techniques have demonstrated significant effectiveness in reducing runoff and peak discharge. However, optimizing their design remains challenging. Dadrasajirlou et al. (this book) propose a new method to find the best percentage cover of LIDs used in urban areas to mitigate floods in urban areas based on machine learning algorithms. The study shows that a combination of different types of LID can be more effective in flood hazard mitigation in urban areas than single types of LID.

Unger et al.'s (this book) study on flood mitigation strategies in Ljubljana's Glinščica River catchment contributes significantly to SDG 11 and SDG 13 by evaluating a range of flood mitigation measures, including NbS, hybrid, and gray infrastructure. By focusing on green infrastructure like rain gardens, urban trees, and green roofs, the research highlights their role in enhancing urban resilience against floods and improving water quality. These NbS not only reduce peak discharge and outflow volumes during flood events but also contribute to climate adaptation by regulating urban microclimates and mitigating urban heat island effects. This approach aligns with SDG 11's objective to build sustainable and resilient cities and communities, while also supporting SDG 13 by promoting climate resilience and reducing greenhouse gas emissions through sustainable urban water management practices.

Fernandez and Pan's (this book) first chapter emphasizes the pivotal role of NbS in bolstering urban resilience and advancing SDGs 11 (Sustainable Cities and Communities) and 13 (Climate Action) within the Netherlands' context. Through a comprehensive literature review and policy document analysis, NbS are identified as critical tools for mitigating climate hazards and addressing socioeconomic vulnerabilities while enhancing ecological sustainability. Despite the recognition of resilience as a priority across Dutch municipalities, the analysis reveals a need for more robust and integrated resilience planning frameworks. By promoting proactive policy development that integrates NbS into urban planning strategies, cities can bolster their adaptive capacity to climate change impacts and foster sustainable urban development. This approach not only supports efforts toward climate mitigation and adaptation but also contributes to creating resilient, livable urban environments aligned with global sustainability goals.

The other chapter by Page et al. (this book) focuses on urban carbon emissions reductions through one of the NbS instruments—carbon sequestration. Through a case study in Stockholm County, Sweden, this chapter by Page et al. (this book) investigates changes in carbon sequestration by forest ecosystems over time, and how forests can be managed as NbS for increased carbon sequestration and storage in the face of climate change. It aligns with SDG 11 by examining how urban and peri-urban forests contribute to sustainable city environments, enhancing green spaces, and helping to offset urban carbon emissions. This also addresses SDG 13 by exploring how effective forest management can significantly enhance carbon capture, thereby supporting global climate action efforts. Through a detailed analysis of tree species composition, forest age, and local hydroclimatic conditions, the chapter provides insights into optimizing forest management practices to maximize carbon sequestration, offering a natural solution to mitigate climate change while contributing to Stockholm's goal of net-zero emissions by 2040.

4 NbS and policy action, strong institution

NbS are increasingly recognized as pivotal strategies to enhance both environmental and social well-being. This concept is gaining prominence in governmental and nongovernmental policies and programs (Galecka-Drozda & Bednorz, 2022). Beyond policy frameworks, NbS has seen significant advancement through research and innovation projects, often conducted in real-life labs and supported by initiatives like those of the European Commission and various academic studies (Zingraff-Hamed et al., 2021). However, integrating NbS into urban governance and municipal agendas remains a significant challenge. Identified hurdles include the risk of NbS remaining vaguely defined without operational rigor (Nesshöver et al., 2017), insufficient stakeholder engagement and local contextualization (Kauark-Fontes et al., 2023), and the persistence of siloed governmental structures that hinder the mainstream adoption of NbS in planning and governance cultures (Dorst et al., 2022).

In addition to implementation and scaling challenges, policy practices for NbS are intricately linked to several other SDGs, such as SDG 1 (No Poverty), SDG 4 (Quality Education), SDG 5 (Gender Equality), SDG 10 (Reduced Inequalities), SDG 16 (Peace, Justice, and Strong Institutions), and SDG 17 (Partnership for Goals). While these connections may not always be explicitly discussed in NbS literature, inclusive, participatory, and collaborative policies and governance for NbS play crucial roles in advancing these broader goals. This book delves into policy actions for NbS across three chapters, offering implications for a range of interconnected SDGs.

The chapter from Scolobig et al. (this book) emphasizes the urgent need to integrate NbS into governance frameworks at local, regional, and national levels to achieve significant environmental and socioeconomic benefits. It identifies key barriers to NbS implementation, such as stakeholder conflicts and knowledge gaps, and proposes recommendations to address these challenges. These include promoting mandatory policy instruments, unlocking public and private funding, and strengthening capacity building and knowledge sharing. By advocating for inclusive governance and innovative financing mechanisms, the chapter supports SDGs such as No Poverty (SDG 1), Reduced Inequalities (SDG 10), and Partnership for Goals (SDG 17). The chapter builds on the results of extensive stakeholder deliberations involving over 70 NbS experts and knowledgeable stakeholders at the national, European, and international scales over 4 years.

The chapter from Cong and Temkin (this book) emphasizes the critical need for effective policy communication to ensure broad understanding and support for NbS among stakeholders. This involves promoting equitable dialog, knowledge sharing, and inclusive decision-making processes, aligning with SDGs like SDG 17 (Partnership for Goals) by advocating for collaborations across sectors. By integrating NbS into local spatial planning and fostering educational efforts (SDG 4—Quality Education), the chapter introduces a support tool designed to integrate NbS into local spatial planning processes, aiding local authorities and urban planning offices, especially during the early planning phases, to collectively identify suitable options.

The second chapter of Fernandez and Pan (this book) delves into integrating NbS and urban resilience into governance and policy agendas, highlighting the imperative of breaking down silos within city governments. It underscores the need for collaborative governance across scales to achieve transformative change, examining drivers of policy adoption such as internal factors (like problem pressure) and external factors (such as network participation). Despite challenges, cities facing significant stressors show a greater inclination toward adopting resilience policies, emphasizing the importance of addressing urban vulnerabilities amidst climate uncertainties. The chapter advocates for multinational support to overcome institutional barriers and proposes incentivizing resilience adoption at national and global levels to ensure all cities benefit from resilient infrastructure and policies. This aligns with SDG 11 (Sustainable Cities and Communities) by promoting inclusive, safe, resilient, and sustainable urban development, and with SDG 17 (Partnership for Goals) through fostering partnerships for effective policy diffusion. It also relates to SDG 16 (Peace, Justice, and Strong Institutions) by advocating for inclusive decision-making processes, and to SDG 1 (No Poverty), SDG 5 (Gender Equality), and SDG 10 (Reduced Inequalities) by aiming for equitable access to resilient solutions that benefit all urban residents, particularly marginalized communities.

5 NbS and other SDGs

NbS contribute significantly to various SDGs, although this book cannot comprehensively cover all connections due to its scope limitations. This section will highlight some important connections between NbS and SDGs that are not extensively discussed in the book. Key SDGs include SDG 6 (Clean Water and Sanitation) and SDG 14 (Life Below Water), which are crucially linked to green infrastructure and SDG 9 (Industry, Innovation, and Infrastructure), as well as the circular economy and SDG 12 (Responsible Consumption and Production). Additionally, this section will explore potential future SDGs that NbS could support, along with the interconnectedness between multiple SDGs.

Among the most critical connections discussed is NbS's role in addressing water-related SDGs, particularly SDG 6 (Clean Water and Sanitation) and SDG 14 (Life Below Water). Freshwater resources, essential for economic activities and daily use, often suffer from exploitation that harms natural ecosystems. Human actions, such as changes in land use, excessive water extraction, and pollution, have significantly impacted around 70% of the world's rivers, posing threats to aquatic biodiversity and global development (Boelee et al., 2017; WWAP, 2015). Healthy freshwater ecosystems, including rivers, lakes, wetlands, and aquifers, are critical for biodiversity conservation and provide vital services such as water purification, storage, and flood mitigation.

NbS, such as utilizing natural wetlands for water storage, provide a holistic method for harmonizing biodiversity conservation and ecosystem service delivery with water management objectives. This approach is in line with an integrated ecosystem strategy that aims for sustainable management of land, water, and living resources, thereby promoting conservation and sustainable utilization practices. By embedding biodiversity considerations into the policies and operations of major public and private stakeholders, the effectiveness of these solutions can be significantly enhanced. Integrated implementation of NbS can optimize multiple benefits and enhance the efficacy of water management strategies (Liu et al., 2023).

NbS are intricately linked to SDG 9 (Industry, Innovation, and Infrastructure), embodying their definition through green infrastructure that offers effective, cost-efficient, and multifunctional alternatives to traditional engineering methods. Green infrastructure underscores the importance of interconnected green spaces, blending natural and human-made elements as essential components of urban development strategies (Fang et al., 2023). NbS, encompassing a broader concept, addresses urban environmental challenges through a problem-solving approach that integrates urban nature into solutions. Historically, technical responses like air conditioning during heatwaves and flood-control dams have dominated urban environmental management (Kim & Kang, 2023). In contrast, green infrastructure and NbS present practical frameworks that enhance urban resilience; examples include green walls and vegetated drainage basins that improve air quality, manage stormwater, and bolster urban resilience against environmental stresses (Ronchi et al., 2020). Collectively, NbS and GI frameworks contribute to robust urban infrastructure aligned with SDG 9, promoting innovation and sustainability in urban planning and development (Blackwood et al., 2022).

Moreover, NbS and green infrastructure also support SDG 12 (Decent Work and Economic Growth) by advancing the principles of a circular economy, which seeks to minimize waste and optimize resource use. This symbiotic relationship illustrates the interconnectedness of SDGs that NbS can foster. NbS solutions align with circular economy principles—such as reusing, reducing, recycling, remanufacturing, and repurposing—thereby closing material and energy loops and minimizing waste generation (Pearlmutter et al., 2020). This approach contrasts with traditional gray infrastructure reliant on nonrenewable materials like concrete and asphalt, which increase impervious surfaces, exacerbate urban heat island effects, disrupt natural water cycles, and adversely affect human well-being due to a lack of urban green spaces (Stefanakis et al., 2021). By promoting cyclical and regenerative innovations in societal production and consumption practices, NbS play a pivotal role in achieving sustainable and circular economic models (Prieto-Sandoval et al., 2018).

NbS have the potential to support numerous other SDGs, even though explicit discussions in the literature might be lacking for some connections. It is essential to recognize that established NbS evidence often supports a nexus of multiple SDGs. While NbS are predominantly evaluated for their water management benefits, they are also crucial for addressing the Water-Energy-Food (W-E-F) nexus and promoting circular economy principles (Carvalho et al., 2022). NbS designed for water management can mitigate trade-offs and simultaneously strengthen water, energy, and food security (Langergraber et al., 2020).

Other examples include that NbS enhance life systems through nutrient cycling and primary production, thereby supporting SDGs 3 (Good Health and Well-being), 6 (Clean Water and Sanitation), 12 (Responsible Consumption and Production), and 15 (Life on Land). They improve food security, water quality, and renewable energy, aligning with SDGs 2 (Zero Hunger), 3 (Good Health and Well-being), 6 (Clean Water and Sanitation), 7 (Affordable and Clean Energy), and 12 (Responsible Consumption and Production). Additionally, NbS regulate climate and water cycles, contributing to SDGs 11 (Sustainable Cities and Communities) and 13 (Climate Action). Beyond environmental benefits, NbS enrich cultural values, supporting SDGs 1 (No Poverty), 4 (Quality Education), 15 (Life on Land), and 16 (Peace, Justice, and Strong Institutions). They also boost urban economies, relevant to SDGs 1 (No Poverty), 8 (Decent Work and Economic Growth), and 9 (Industry, Innovation, and Infrastructure) (Schmidt et al., 2022). They contribute to environmental health by filtering pollutants and reducing reliance on energy-intensive treatments, as demonstrated by healthy forests, wetlands, and sustainable agriculture practices (Keestra et al., 2023).

References

Blackwood, L., Renaud, F. G., & Gillespie, S. (2022). Nature-based solutions as climate change adaptation measures for rail infrastructure. *Nature-Based Solutions, 2*, 100013.

Boelee, E., Janse, J., Le Gal, A., Kok, M., Alkemade, R., & Ligtvoet, W. (2017). Overcoming water challenges through nature-based solutions. *Water Policy, 19*(5), 820–836.

Carvalho, P. N., Finger, D. C., Masi, F., Cipolletta, G., Oral, H. V., Tóth, A., … Exposito, A. (2022). Nature-based solutions addressing the water-energy-food nexus: Review of theoretical concepts and urban case studies. *Journal of Cleaner Production, 338*, 130652.

Cortinovis, C., Olsson, P., Boke-Olén, N., & Hedlund, K. (2022). Scaling up nature-based solutions for climate-change adaptation: Potential and benefits in three European cities. *Urban Forestry & Urban Greening, 67*, 127450.

Dorst, H., van der Jagt, A., Toxopeus, H., Tozer, L., Raven, R., & Runhaar, H. (2022). What's behind the barriers? Uncovering structural conditions working against urban nature-based solutions. *Landscape and Urban Planning, 220*, 104335. https://doi.org/10.1016/j.landurbplan.2021.104335.

European Commission. (2022). *Commission announces 100 cities participating in EU Mission for climate-neutral and smart cities by 2030.*

Fang, X., Li, J., & Ma, Q. (2023). Integrating green infrastructure, ecosystem services and nature-based solutions for urban sustainability: A comprehensive literature review. *Sustainable Cities and Society*, 104843.

FAO. (2015). *Growing greener cities in Africa; First status report on urban and peri-urban horticulture in Africa.* Food and Agriculture Organization of the United Nations: Rome, Italy.

Ferreira, C. S. S., Duarte, A. C., Boulet, A. K., Veiga, A., Maneas, G., & Kalantari, Z. (2023). Agricultural land degradation in Portugal and Greece. In P. Pereira, M. Muñoz-Rojas, I. Bogunovic, & P. Panagos (Eds.), *Impact of agriculture on soil degradation II. The handbook of environmental chemistry.* Berlin, Heidelberg: Springer. https://doi.org/10.1007/698_2022_950.

Ferreira, C. S. S., Seifollahi-Aghmiuni, S., Destouni, G., Ghajarnia, N., & Kalantari, Z. (2022). Soil degradation in the European Mediterranean region: Processes, status and consequences. *Science of the Total Environment*, 805, 150106. https://doi.org/10.1016/j.scitotenv.2021.150106.

Galecka-Drozda, A., & Bednorz, L. (2022). Outside the forest stand: Analysis of urban forest buffer zones to implement nature-based solutions—A case study of Poznań (Poland). *Sylwan*, *166*(05).

Gonzalez-Flo, E., Romero, X., & García, J. (2023). Nature based-solutions for water reuse: 20 years of performance evaluation of a full-scale constructed wetland system. *Ecological Engineering*, *188*, 106876. https://doi.org/10.1016/j.ecoleng.2022.106876.

IPCC. (2023). In Core Writing Team, H. Lee, & J. Romero (Eds.), *Climate change 2023: AR6 synthesis report. Contribution of working groups I, II and III to the sixth assessment report of the intergovernmental panel on climate change.* Geneva: IPCC. https://www.ipcc.ch/report/sixth-assessment-report-cycle/.

James, P., Banay, R. F., Hart, J. E., & Laden, F. (2015). A review of the health benefits of greenness. *Current Epidemiology Reports*, *2*(2), 131–142.

Kalantari, Z., Ferreira, C. S. S., Pan, H., & Pereira, P. (2023). Nature-based solutions to global environmental challenges. *Science of the Total Environment*, *880*, 163227. https://doi.org/10.1016/j.scitotenv.2023.163227.

Kauark-Fontes, B., Marchetti, L., & Salbitano, F. (2023). Integration of nature-based solutions (NBS) in local policy and planning toward transformative change. Evidence from Barcelona, Lisbon, and Turin. *Ecology and Society*, *28*(2).

Keestra, S., Veraart, J., Verhagen, J., Visser, S., Kragt, M., Linderhof, V., … Groot, A. (2023). Nature-based solutions as building blocks for the transition towards sustainable climate-resilient food systems. *Sustainability*, *15*(5), 4475. https://doi.org/10.3390/su15054475.

Kim, J., & Kang, J. (2023). Development of hazard capacity factor design model for net-zero: Evaluation of the flood adaptation effects considering green-gray infrastructure interaction. *Sustainable Cities and Society*, *96*, 104625.

Kunming-Montreal Global Biodiversity Framework. (2022). *Conference of the parties to the convention on biological diversity.* https://www.cbd.int/doc/c/e6d3/cd1d/daf663719a03902a9b116c34/cop-15-l-25-en.pdf (Accessed 14 February 2024).

Langergraber, G., Pucher, B., Simperler, L., Kisser, J., Katsou, E., Buehler, D., … Atanasova, N. (2020). Implementing nature-based solutions for creating a resourceful circular city. *Blue-Green Systems*, *2*(1), 173–185.

Liu, L., Dobson, B., & Mijic, A. (2023). Optimisation of urban-rural nature-based solutions for integrated catchment water management. *Journal of Environmental Management*, *329*, 117045.

Maes, J., & Jacobs, S. (2017). Nature-based solutions for Europe's sustainable development. *Conservation Letters*, *10*, 121–124.

Martín, E. G., Giordano, R., Pagano, A., Van Der Keur, P., & Costa, M. M. (2020). Using a system thinking approach to assess the contribution of nature based solutions to sustainable development goals. *Science of the Total Environment, 738,* 139693.

Nesshöver, C., Assmuth, T., Irvine, K. N., Rusch, G. M., Waylen, K. A., Delbaere, B., Haase, D., et al. (2017). The science, policy and practice of nature-based solutions: An interdisciplinary perspective. *Science of the Total Environment, 579,* 1215–1227.

Newell, J. P., Seymour, M., Yee, T., Renteria, J., Longcore, T., Wolch, J. R., & Shishkovsky, A. (2013). Green alley programs: Planning for a sustainable urban infrastructure? *Cities, 1,* 144–155. https://doi.org/10.1016/j.cities.2012.07.004.

Pan, H., Page, J., Shi, R., Cong, C., Cai, Z., Barthel, S., ... Kalantari, Z. (2023). Contribution of prioritized urban nature-based solutions allocation to carbon neutrality. *Nature Climate Change, 13*(8), 862–870.

Pearlmutter, D., Theochari, D., Nehls, T., Pinho, P., Piro, P., Korolova, A., ... Pucher, B. (2020). Enhancing the circular economy with nature-based solutions in the built urban environment: Green building materials, systems and sites. *Blue-Green Systems, 2*(1), 46–72.

Pereira, P., Ferreira, C. S. S., & Zhao, W. (2024). Editorial: Nature-based solutions for ecosystem restoration. *Current Opinion in Environmental Science & Health, 39,* 100546. https://doi.org/10.1016/j.coesh.2024.100546.

Potocki, K., Raska, P., Ferreira, C. S. S., & Bezak, N. (2023). Translating nature-based solutions for water resources management to higher educational programs in three European countries. *Land, 12,* 2050. https://doi.org/10.3390/land12112050.

Prieto-Sandoval, V., Jaca, C., & Ormazabal, M. (2018). Towards a consensus on the circular economy. *Journal of Cleaner Production, 179,* 605–615.

Ronchi, S., Arcidiacono, A., & Pogliani, L. (2020). Integrating green infrastructure into spatial planning regulations to improve the performance of urban ecosystems. Insights from an Italian case study. *Sustainable Cities and Society, 53,* 101907.

Sachs, J., Lafortune, G., & Fuller, G. (2024). *The SDGs and the UN summit of the future. Sustainable development report 2024.* Paris, Dublin: SDSN, Dublin University Press. https://doi.org/10.25546/108572.

Schmidt, S., Guerrero, P., & Albert, C. (2022). Advancing sustainable development goals with localised nature-based solutions: Opportunity spaces in the Lahn river landscape, Germany. *Journal of Environmental Management, 309,* 114696.

Seddon, N., et al. (2020). Understanding the value and limits of nature-based solutions to climate change and other global challenges. *Philosophical Transactions of the Royal Society B, 375,* 20190120.

Shukla, P. R., Skea, J., Calvo Buendia, E., et al. (2019). *Climate change and land: An IPCC special report on climate change, desertification, land degradation, sustainable land management, food security, and greenhouse gas fluxes in terrestrial ecosystems.* Intergovernmental Panel on Climate Change.

Sonneveld, B., Merbis, M., Alfarra, A., Ünver, O., & Arnal, M. (2018). *Nature-based solutions for agricultural water management and food security* (p. 68). Rome: FAO. Land & Water Discussion Paper no 12.

Stefanakis, A. I., Calheiros, C. S., & Nikolaou, I. (2021). Nature-based solutions as a tool in the new circular economic model for climate change adaptation. *Circular Economy and Sustainability, 1,* 303–318.

Tate, C., Wang, R., Akaraci, S., Burns, C., Garcia, L., Clarke, M., & Hunter, R. (2024). The contribution of urban green and blue spaces to the United Nation's sustainable development goals: An evidence gap map. *Cities, 145,* 104706.

UNESCO. (2021). *The United Nations world water development report 2021: Valuing water.* UNESCO. https://unesdoc.unesco.org/ark:/48223/pf0000375724/PDF/375724eng.pdf.multi.

United Nations General Assembly. (2015). *Transforming our world: The 2030 agenda for sustainable development.*

WWAP. (2015). *The United Nations world water development report 2015: Water for a sustainable world.* Paris: UNESCO.

Zhang, L., Cong, C., Pan, H., Cai, Z., Cvetkovic, V., & Deal, B. (2021). Socioecological informed comparative modeling to promote sustainable urban policy transitions: Case study in Chicago and Stockholm. *Journal of Cleaner Production, 281,* 125050.

Zingraff-Hamed, A., Hüesker, F., Albert, C., Brillinger, M., Huang, J., Lupp, G., ... Schröter, B. (2021). Governance models for nature-based solutions: Seventeen cases from Germany. *Ambio, 50*(8), 1610–1627. https://doi.org/10.1007/s13280-020-01412-x.

Section 1

NbS to support sustainable development goals

Chapter 1.1

Forests for climate change mitigation: Temporal dynamics of carbon sequestration in the forests of Stockholm County

Jessica Page[a], Victoria Kareflod[b], and Elisie Kåresdotter[b]

[a]*Department of Physical Geography, Stockholm University, Stockholm, Sweden,* [b]*Department of Sustainable Development, Environmental Science and Engineering (SEED), School of Architecture and the Built Environment (ABE), KTH Royal Institute of Technology, Stockholm, Sweden*

1 Introduction

Carbon sequestration and storage will be a crucial complement to reducing anthropogenic greenhouse gas (GHG) emissions in mitigating climate change in the coming decades (IPCC, 2023). Globally, forests are seen as a valuable natural solution to climate change due to their significant carbon sequestration and storage potential (Burke et al., 2021). They are relied upon both locally and internationally to offset GHG emissions as countries, cities, and even private companies strive for "net zero" emissions. While man-made carbon capture and storage can be costly and are often met with public skepticism, forests offer a range of ecosystem services and economic benefits in addition to carbon sequestration (Hansen & Malmaeus, 2016; Lane et al., 2021).

The forest biome is important for natural climate regulation (Dar et al., 2020). However, there is significant uncertainty associated with attempting to quantify how much carbon forests can capture and store, and they are often overlooked in carbon accounting strategies (Kuittinen et al., 2016). Much of this uncertainty stems from the great variation observed in sequestration capacity across different forests. Different tree species grow and sequester carbon at different rates, which vary over the tree's lifetime (Bradford & Kastendick, 2010). Sequestration capacity also varies with species combinations, geographic location, and over time according to the local hydroclimatic conditions

Nature-Based Solutions in Supporting Sustainable Development Goals
https://doi.org/10.1016/B978-0-443-21782-1.00001-4

(Li et al., 2015; Zhu et al., 2019). Research on vegetation age has been conducted across various regions, examining specific species at different latitudes over their lifetimes. These studies consider factors such as soil, debris, biomass, and other relevant criteria. Soil plays a crucial role in carbon accumulation, storing significant amounts of carbon, and is important in evaluating a carbon sink's potential. Once soils reach a steady state, the accumulation rates might decline, although this process can take an exceptionally long time (De Simon et al., 2012; Ostrogović Sever et al., 2019; Uri et al., 2012). Although forests are generally accepted to be significant carbon sinks over their lifetime, some studies have found that carbon sequestration slows as the forest ages and even that old-growth forests can eventually become sources of carbon to the environment on an annual basis (Petersson et al., 2022). For example, a study of a Swedish forest, left largely undisturbed since 1796, found this forest to be a net source of carbon to the atmosphere over the study period of 10 years (Hadden & Grelle, 2017).

Understanding the spatiotemporal variation of carbon sequestration in forests will be important in the coming decades as countries seek to achieve carbon neutrality and net zero climate action goals. While the study of long-term carbon sequestration dynamics in vegetation is a rapidly advancing area of research, the majority of studies still focus on short-term carbon sequestration, and scaling these findings up to a long-term context remains a challenge, and more research is needed in this field (Lal, 2007; Lorenz & Lal, 2010). This is particularly relevant in regions with diverse landscapes, such as forests, wetlands, agricultural areas, and parks, which offer unique opportunities for sustainable management practices to enhance carbon sequestration (Colding, 2013).

Despite widespread agreement that forests are valuable and important as carbon sinks (and for their many other environmental and social benefits), forests are under threat in many places around the world (d'Annunzio et al., 2015; Hansen et al., 2013; Karger et al., 2021). One of the major causes of deforestation globally is land use change, in which forest is removed to make way for human activities such as agriculture or urban expansion (Hosonuma et al., 2012). This is problematic for climate change mitigation efforts on two fronts: not only is the future carbon sequestration capacity lost, but the deforested land often also becomes a carbon source as the trees are removed and the soil is disturbed (Guo & Gifford, 2002). Climate change itself also poses a threat to many forests through increasingly frequent and severe natural hazards like floods, droughts, and forest fires. Changing hydroclimatic conditions may make large areas unsuitable for many of the plant species in existing forests, while also making them more suitable for invasive species and pests (Lindner et al., 2014).

In this chapter, we investigate how the carbon sequestration potential of forests in Stockholm County, Sweden, is expected to change with the aging of the forests in the coming decades and discuss how this, together with other factors such as climate change and land use change, could impact their contributions to achieving carbon neutrality goals.

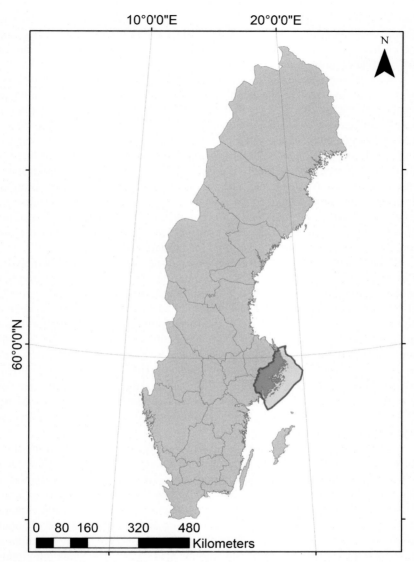

FIG. 1 Map showing the location of Stockholm County in Sweden, with the county borders shown in *red*.

1.1 Case study area

Stockholm County, Sweden, was chosen for a case study to assess the long-term dynamics of carbon sequestration by forests over a large area. Fig. 1 shows the location of this county in Sweden. Stockholm County has an area of about 6519 km², approximately 55% of which is forest (SCB, 2023a). The county also includes the Swedish capital city of Stockholm, other smaller cities and towns, farmland, and the many islands of the Stockholm Archipelago. The forests in Stockholm County consist of 11% broad-leaf forest (mainly birch, oak, and beech trees), 76% coniferous (mainly spruce and pine trees), and 13% mixed forests (Naturvårdsverket, 2020).

Forest covers a large part of Sweden and is mainly grown as part of forestry, with only around 15% of Stockholm County forests being forest that is not part of forestry (SCB, 2023a). Despite being grown for economic purposes, these forests are accessible to the public and can be used for recreational purposes. Overall, protected areas account for 12% of Stockholm County, mainly consisting of nature reserves (SCB, 2023b). The majority of the forest in Stockholm is therefore not protected, but there is a law (Skogsvårdslag 1979:429) that states that new trees must be planted after harvesting, allowing new sequestration. Currently, the mean annual forest growth is higher than the harvest, meaning a net growth. However, the trend is toward increased harvest, meaning that this could change, especially if the forest growth is negatively affected or if more areas are converted for other purposes (Naturvårdsverket, 2023).

Following the 2015 Paris Agreement, countries have submitted Nationally Determined Contributions (NDCs) and committed to new climate goals to limit global temperature rise to below 2°C, with an aspiration to stay below 1.5°C (UNFCC, 2022). Sweden's NDC aims for net zero carbon dioxide (CO_2) emissions by 2045 (Swedish Ministry of the Environment, 2019). To support these goals, Stockholm City developed a Climate Action Plan (CAP) in 2016, outlining concrete strategies to reduce greenhouse gas emissions by 19 million tons and achieve fossil-free, zero-emissions status by 2040 (Stockholms Stad, 2016). Region Stockholm also has a CAP that includes the entire county and not just the capital city, in which they aim to achieve net zero emissions by 2045, in line with the Paris Agreement's neutrality goal of 2045 (Tillväxt- och Regionplaneförvaltningen, 2019).

Stockholm City's CAP includes strategies to reduce emissions from transport, heating and cooling in buildings, and electricity and gas production (Stockholms Stad, 2016). To achieve net zero goals, compensatory measures for remaining emissions and acceleration of emissions reduction are essential. Stockholm City has invested in expensive offset measures like carbon capture and storage and bioenergy with carbon capture and storage. These cost about 1000 SEK per ton of CO_2 captured, with a projected need to capture 1,300,000 tons of CO_2 annually (Stockholms Stad, 2020). The Stockholm roadmap also explores alternative carbon sequestration approaches, highlighting the potential of green areas to enhance natural ecosystems' capacity to absorb atmospheric carbon dioxide (Tillväxt- och Regionplaneförvaltningen, 2019). While ecosystems like forests and wetlands can capture carbon, their potential in the county is unclear and not fully accounted for in local carbon accounting strategies, although large carbon sinks are included in national GHG emission accounting.

Stockholm is often touted as a global front-runner city and region for climate action and is part of programs such as the European Commission's Net Zero Cities (Net Zero Cities, 2022). However, despite their ambitious climate action plans and a good deal of investment in emissions reduction across many sectors, Stockholm will very likely still rely on the sequestration provided by the county's vast forests in order to achieve net zero greenhouse gas emissions goals (Page et al., 2021).

2   Methods and data

2.1   Forest species and age distribution

Land cover data from Svenska Marktäckedata (Naturvårdsverket, 2020) was combined with the most recent available detailed forest data from the Swedish University of Agricultural Sciences (SLU) to derive the volume distribution of tree species within the boundary of Stockholm County (Lavalle et al., 2015; Swedish University of Agricultural Sciences, 2015). The difference between total forest volume and the combined volumes of the five most common tree species represents the volume of other tree species present in the forests, which we have classified for the purposes of this study as "other". These species volumes were converted into area coverage based on ratios obtained by processing the data using GIS software.

Data on tree species age from SLU were used together with data on Swedish forest harvesting and renewal practices (Naturvårdsverket, 2023; Swedish University of Agricultural Sciences, 2015) to approximate the age distribution of the forests in Stockholm County for the base year of 2015. The ratios for species within age classifications were derived from SLU's forest age classification system (Swedish University of Agricultural Sciences, 2017). Two of the SLU tree age classes were merged to form the middle-aged category to align with the classifications prevalent in the broader literature on forest carbon sequestration. Further data from SLU on species distribution within these age classes provided the necessary ratios of tree species across the three age classes for the calculation of the area per species, forest types, and age class for the 2015 scenario (Swedish University of Agricultural Sciences, 2022). A summary of the spatial data used for forest mapping and calculations is provided in Table 1.

To simulate an aging forest, we assumed a continuous distribution of tree ages within each age class. The total share of a species within an age class was divided by the total number of years included in that class. The 25-year study period was used to multiply the annual volume of species per year to reflect forest aging. Annual tree mortality (including harvesting) was then calculated and subtracted from the relevant age classes. The annual volume of dead wood (3.5 million m^3) (Naturvårdsverket, 2023) provided a mortality ratio relative to the total volume, which was used to estimate the species' total coverage in 2040. This ratio, reflecting the volume decrease due to mortality, was then

TABLE 1 Spatial data used for forest mapping and calculations.

Data description	Data name	Source
Forest species distribution	Skogsdata—volumes by species	Swedish University of Agricultural Sciences (2015)
Forest age distribution	Skogsdata—tree age	Swedish University of Agricultural Sciences (2015)
Land use	Svenska Marktäckdata 2000	Naturvårdsverket (2020)
Stockholm County Boundary	UI—Boundaries for the functional urban areas	Lavalle et al. (2015)

applied to calculate a new representative area for subsequent calculations. The following formula was used to simulate the change in area for each tree species based on the annual mortality rate:

$$\text{Area}_{tot} \ (1 - \text{mortality rate})^{\text{passing years}}$$

The mortality rate, sourced from the Swedish University of Agricultural Sciences (2017) and applicable to the Svealand forest, was projected onto Stockholm County Forest, which is part of Svealand. Mortality projections were distributed among middle-aged and mature trees based on forest management practices that restrict tree cutting until ages 45–100 for coniferous trees, age 35 for birch, and ages 80 and 90 for beech and oak, respectively (Skogsstyrelsen, 2024). A simplification was made by estimating total mortality to be equally distributed among species. Finally, tree renewal rates (including replanting practices) for Stockholm forests were estimated based on data from the Swedish University of Agricultural Sciences (2017) and added to the young forest category.

2.2 Literature search

A literature search was conducted in order to establish the sequestration potential associated with each tree species commonly found in Stockholm's forests across three age ranges. The search was limited to peer-reviewed studies and government documents and focused on those in which the net ecosystem productivity was measured using chronosequence and eddy covariance approaches. Other measurement approaches were included when this was not available. Net ecosystem productivity (NEP), defined as the difference between organic carbon (OC) in gross primary production (GPP) and net ecosystem respiration (NER), represents the sum of available OC storage within the system (Lovett et al., 2006). The magnitude of NEP indicates the carbon sequestration rate, where positive values denote a carbon sink and negative values indicate a carbon source. This measurement method was selected as it accounts for the above- and below-ground vegetation elements as well as the important role of the soil in carbon sequestration (Clay et al., 2020). The holistic and dynamic nature of NEP is valuable for evaluating the sequestration potential of large areas over extended periods of time and through multiple life stages of the vegetation.

The chronosequence approach used in this study evaluates long-term changes in NEP by considering species dynamics and terrestrial biomes of varying ages through field measurements. This widely used method assesses stand development dynamics and effectively estimates carbon budgets and dynamics on a larger scale in a relatively short time, essential for effective environmental management and policy design (Uri et al., 2012). This approach represents different aged sites as points in time within their development stages, studying them in parallel to account for temporal changes using a "space-for-time" technique. It is commonly applied in ecosystem sciences to assess nutrient accumulation dynamics (Yanai et al., 2003).

Similar to the chronosequence approach, eddy covariance can be employed to assess NEP, accounting for temporal and spatial patterns while simultaneously analyzing additional indicators such as mean annual precipitation and temperature (Liang & Wang, 2020). Eddy covariance is an advanced remote sensing technology used for direct observation of gas exchanges between ecosystems and the atmosphere. It is increasingly utilized in CO_2 analysis due to its capability for precise data acquisition over both short- and long-term periods, ranging from hours to years. This method is applied in many ecosystems globally today, measuring carbon cycle interactions within the atmospheric and terrestrial biosphere networks and fluxes.

Little of the available literature includes studies conducted within Stockholm County, but priority was given to studies of vegetation located as close as possible to Stockholm, followed by studies conducted in regions with similar climatic conditions, and the data completed with literature from further afield. When possible, several studies on the same species were used to get an average of the values in the classes. The age classifications young, middle-aged, and mature were used to identify the current age distribution of the tree species present in Stockholm's forests.

Several limitations were encountered in the literature search, with no single values available for some of the tree species for some of the age classes. No NEP values were found for young beech trees through the literature search; these were therefore projected from the middle-aged age class. The representation of young beech in the study area is exceptionally low and therefore had minimal impact on the result. No single NEP value could be found for mature birch trees, so an average NEP was estimated based on several studies featuring old-growth forests with significant representation of birch trees. Although birch is the dominant species in broad-leaf forests, mature birch trees represent a very small proportion of the total forest, so this also had a minimal impact on the overall results. However, the long-term sequestration potential of old-growth forests, including birch trees, is relevant for the management of forests in Stockholm and elsewhere, so this uncertainty is discussed further in subsequent portions of this chapter. As no data were available concerning the species of trees included in the "other" category, an average value of the NEP of the most common trees for each age class was used to estimate sequestration for this category.

3 Results

3.1 Forest species distribution in Stockholm County in 2015–40

It was calculated that in 2015, Stockholm County had a total forest area of 345270.32 ha. The coniferous forest was dominant in the region, covering 76.01% of the total forest area (262445.24 ha), broad-leaf forests covered 11.03% of the forest area (38093.12 ha), and mixed forest made up 12.96% of all forest total area (44731.97 ha). The spatial distribution of the five main tree species in these forests (beech, birch, oak, pine, and spruce) across Stockholm County in 2015 is shown in Fig. 2.

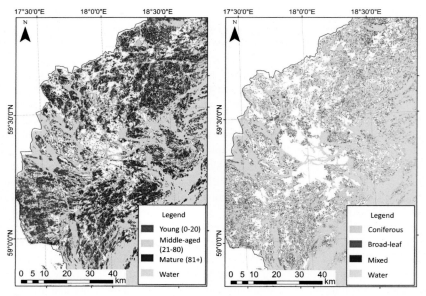

FIG. 2 The distribution of land classified as different forest types across Stockholm County for the year 2015 (left) and tree age distribution for the year 2010 (right). *(Based on Naturvårdsverket, (2020). Svenska Marktäckedata 2000. Swedish University of Agricultural Sciences, (2015). SLU Skogskarta. [WWW Document]. SLU.SE. URL https://www.slu.se/centrumbildningar-och-projekt/ riksskogstaxeringen/statistik-om-skog/slu-skogskarta/ (Accessed 30 May 2024).*

The 2040 estimations used mortality and renewal of forests to calculate the changes in the composition of the forest types over a period of 25 years. By 2040, the area of coniferous forests is projected to have increased to 287848.05 ha, broad-leaf forest increased to an estimated 47997.78 ha, and mixed forest to cover 52669.76 ha. It should be noted that these results represent only the changes due to forest mortality and renewal and do not include the impacts of human-driven land use change such as urban expansion or land management practices to prevent the natural spread of the forest over time.

3.2   Forest age dynamics

The spatial distribution of forest age across Stockholm County for the year 2010 (the most recent available forest age data) is shown in Fig. 2. The majority of the forests in Stockholm County are dominated by trees in the young and middle-aged classes, with some old-growth forests where mature trees are dominant located in the southeast of the county. The projected change in age distribution between 2015 and 2040 is shown for each tree species in Fig. 3. Here it can be seen that the age distribution is expected to shift from a dominance of young and middle-aged trees to middle-aged and mature trees, based on current management and harvesting practices.

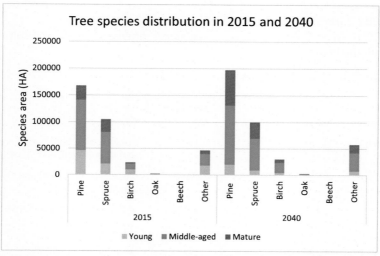

FIG. 3   Projected changes in forest age and species distribution from 2015 to 2040.

3.3 Forest NEP trends

The forests follow a trend showing a lower sequestration potential in the young age class with a considerable increase in the middle-age class, where the potential peaks, followed by a decrease at mature age. Although much of the available literature finds that old-growth forests continue to sequester carbon (Luyssaert et al., 2008), several studies suggest that while the total carbon stocks remain high, the annual NEP of mature forest declines significantly after middle age to values that are near or even below zero (Bradford & Kastendick, 2010; Taylor et al., 2014; Uri et al., 2012). The results of our literature search suggest that Swedish forests could remain as net carbon sinks or potentially shift to become sources of carbon to the atmosphere as they age, depending on a great many factors, including forest composition and management practices. The NEP values resulting from the literature search are shown in Table 2. None of the NEP values we found for individual tree species in the mature age category were negative, but they do show a trend of decreasing with age, and one study found that the NEP of birch trees on drained peatlands became negative in the mature age range (Uri et al., 2017). This value was not used in our calculations, as our data do not suggest that a significant portion of the forests in Stockholm County are on drained peatlands. However, this finding and the possibility that old-growth forests could eventually become net carbon sources are considered together with the implications of this for forest management in the discussion section of this chapter.

TABLE 2 The NEP values in tons of carbon per hectare per year ($tCHA^{-1}$ $year^{-1}$) for each type of tree across different age ranges resulting from the literature search.

		Age range		
Species	Young (0–20)	Middle age (20–80)	Mature (80+)	Source
Birch	4.38	4.30	0.11	Gough et al. (2008), Taylor et al. (2014), and Varik et al. (2015)
Beech	2.70	2.70	2.77	Bascietto et al. (2004) and De Simon et al. (2012)
Oak	−1.23	2.94	3.08	Ostrogović Sever et al. (2019)
Pine	0.62	2.29	1.58	Uri et al. (2017)
Spruce	−1.90	2.95	1.70	De Simon et al. (2012)
Other	0.47	3.04	1.85	N/A

3.4   Total forest sequestration

When the NEP values are multiplied by the total area of each tree species and age category and converted to carbon dioxide equivalents, we find the total annual sequestration potential for each type of forest in Stockholm. The total annual sequestration potential for the Stockholm County forests in 2015 was calculated to be 2.36 MtCO$_2$-eq. This potential was mostly from middle-aged trees (77%), with smaller contributions by young (7%) and mature trees (15%). The total annual sequestration potential is projected to increase to 2.99 MtCO$_2$-eq in 2040, with a notable increase in the contributions from mature trees (25%) and a slight reduction in the proportion coming from middle-aged trees (75%), although this is still a net increase in sequestration potential from middle-aged trees given the total increase in sequestration. The contribution from young trees is projected to decrease to only 2% of the total by 2040, due to the increasing overall age of the forests. The projected change in sequestration potential of the forests over the 25-year study period and broken down by forest type is shown in Fig. 4.

4   Discussion

Our results project that if forest management and harvesting practices in Stockholm do not change in the coming decades, the sequestration potential offered by the county's forests will increase by around 27% above the 2015 values by 2040. This increase is due to an overall natural expansion of the forest area if left unchecked and also due to the aging of the forest stock over the 25-year study period. This increase could be a boon for the county in its quest for net zero

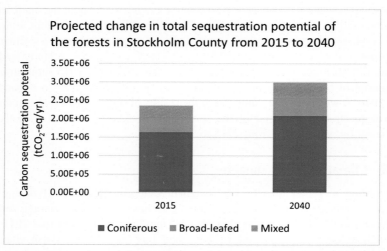

FIG. 4   Graph showing the calculated sequestration by forests in 2015 and projected for 2040.

emissions by 2040, but there are several factors not included in this study as well as a number of possible sources of uncertainty in the assessment of forest carbon sequestration, which could mean that this increase is not realized.

4.1 Limits of forest spread and land use change impacts

Our calculations project an increase in the total forest area in Stockholm County from 345,270 HA in 2015 to 388,516 HA in 2040, which represents the natural spread of the forest over time if it is allowed to grow unchecked. However, this spread is likely to be limited in reality due to human land use, land use change, and land use management practices, which include preventing the spread of forests in certain areas for cultural reasons and preserving other habitats (Eriksson, 2020; Stockholms Stad, 2023). When considering the change in sequestration potential of the forests, limiting the spread of the forest is not likely to have a significant impact over our 25-year study period, as the "spreading" portion of the forest would necessarily be between 0 and 25 years old by 2040. As illustrated in Fig. 5, only a very small proportion of the 2040 sequestration potential is likely to come from the trees in the young age class (0–20 years); indeed, in the dominant coniferous forests, sequestration potential from the trees in this category is close to zero due to the negative NEP associated with young spruce trees.

Although the forests are not likely to spread as significantly as they would if left to grow unchecked, development strategies in Stockholm mean that they are also not expected to shrink significantly due to land use change. While only 11.9% of the forests in Stockholm County are formally protected (SCB, 2023b), the current development strategy in the region has a focus on compact urban forms and preventing sprawl. There is a strong prioritization of green spaces and the inclusion of "green swathes" threaded throughout the main urban region (Tillväxt- och Regionplaneförvaltningen, 2017). If the current regional development plan and practices remain in place, it is unlikely that a significant amount of the existing forest will be lost to land use change. A study by Pan et al. (2020) found that only slightly more than 2% of the 2015 sequestration potential in Stockholm County would likely be lost due to urban expansion by 2040. If we apply this reduction to the results of this study, we still find that the sequestration potential of the forests in Stockholm County is likely to increase by nearly 25% from 2015 to 2040.

4.2 Climate change impacts

According to projections by the Swedish Meteorological and Hydrological Institute, Stockholm is likely to experience an increase in both average annual temperature (1–2°C) and precipitation (2–5 mm/month) in the period 2011–40 (Swedish Meteorological and Hydrological Institute, 2023). These changes could lead to a longer growing season for Swedish forests, and together with

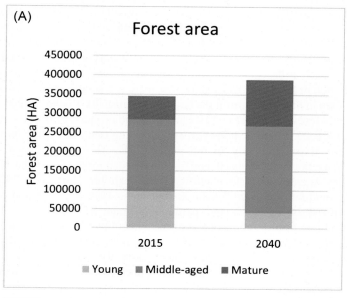

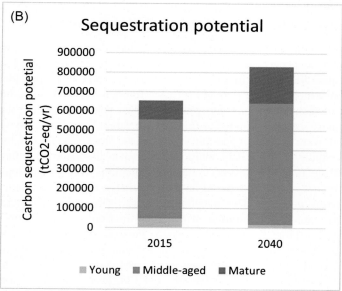

FIG. 5 Graphs showing (A) the change in age composition of Stockholm's forests from 2015 to 2040 and (B) the proportions of sequestration potential attributed to different age classes across the same time period.

the fertilization effects of increasing carbon dioxide, could mean that climate change impacts lead to increased forest growth and sequestration potential beyond that projected in this study (Davis et al., 2022; Felton et al., 2016).

Climate change also brings with it uncertainty and possible increased threats to forests, such as droughts, forest fires, damage from storms, and the spread of pests (Felton et al., 2016). Data show that there is a small annual loss of forest area to fires in Stockholm, with the largest loss since 2001 being 58 HA (0.02% of the total forest area) lost in 2010, and no trend of increase in fire loss in the period 2001–23. There is, however, an increasing trend in the number of fire risk alerts issued annually in the period 2012–23 (Global Forest Watch, 2024). In a report on climate change risks in Sweden for the Swedish Civil Contingencies Agency, the risks identified for Stockholm's forests by 2050 were reduced runoff in the south of the county under both RCP 4.5 and 8.5, and increased risk of insect damage, particularly under RCP 8.5., both of which could decrease sequestration potential in the county (Jaramillo et al., 2021). Overall, the evidence suggests that, although climate change impacts do present increasing risks to the forests in Stockholm, currently it is more likely that these impacts will lead to an increase rather than a decrease in sequestration potential in the county by 2040.

4.3 Uncertainties and limitations in NEP estimations

Because carbon sequestration potential is highly dependent on local factors such as forest species composition and hydroclimatic conditions, it is difficult to generalize NEP values for individual tree species and forest types. Ideally, we would have measured NEP values for each forest in the study area, but in the absence of such data, we have to generalize these values from other studies. In our literature search, we aimed to find and use values from studies in which the conditions were as close as possible to those in the Stockholm forests, but unfortunately, studies with similar conditions were not available for all of the species across all age ranges, so less accurate approximations were used in some cases. This introduces some uncertainty in the calculation of the Stockholm forests' sequestration potential. This uncertainty is seen throughout the literature on vegetative carbon sequestration, with large variations in estimations of both local and global forest carbon sink capacity. For example, different studies have found that old-growth forests can continue to sequester carbon as they age, tend toward a steady state, or become sources of carbon emission (Hadden & Grelle, 2017; Luyssaert et al., 2008; Petersson et al., 2022). The accuracy of the final sequestration values calculated in this study would be greatly improved if more studies measuring tree and forest NEP across different ages were available, and this is an area in which much future research is still needed. Nonetheless, although the values contain some uncertainty, we can still identify trends in the forest composition and sequestration capacity to answer the question of how this sequestration is likely to change in the coming decades and what this means for forests as a nature-based solution to climate change.

4.4 Management of aging forests

As our results show, the sequestration potential of forests changes over time as the trees age. The general trend is that young trees start as either a source of carbon to the atmosphere or with a very small annual sequestration potential. This increases quickly as the trees mature, and annual sequestration potential peaks with a large positive NEP value in middle age (20–80 years) (Grelle et al., 2023). NEP then declines over time as trees reach the mature age group (>80 years) and may eventually tend toward or below zero. It is worth noting here that, although the decrease in NEP may mean that the forest eventually stops sequestering more carbon each year, the cumulative total carbon sequestration and storage by the forest over its lifetime remains very large.

One particularly interesting finding from our literature search is that birch tree stands on drained peatlands became sources of carbon to the atmosphere after middle-age (Uri et al., 2017). This negative NEP value was not used in our calculations, as data were not available to calculate the percentage of mature beech growing on drained peatlands in Stockholm County. However, about 14% of forestry in Sweden is on drained peatlands, so it is possible that at least some of the forest area in Stockholm would fall into that category (Laudon et al., 2023). Our study found some areas of old-growth forests in Stockholm with high concentrations of birch trees; if further investigation into these areas reveals that they are also drained peatlands, local authorities should consider that they might be sources of GHGs to the atmosphere.

In order to maintain and increase the carbon sequestration potential of forests into the future, it may be beneficial to manage them for increased growth and sequestration rather than leaving them undisturbed (Hadden & Grelle, 2017; Petersson et al., 2022). Studies suggest that while thinning may decrease sequestration and make forests more susceptible to pests (Jörgensen et al., 2021; Subramanian et al., 2015), management of forests with regular harvesting for fast-growing, even-aged forests can help to increase carbon sequestration and be protective against potential climate change impacts such as storm damage (Kauppi et al., 2022; Subramanian et al., 2019). Currently, around 85% of the forests in Stockholm are managed for forestry, while just under 12% are protected in nature reserves (SCB, 2023b). Practices in these managed forests in Stockholm should be reviewed as these forests age in order to ensure that they are in line with the latest best practices for carbon sequestration and resilience to climate change impacts, according to research. Future studies are also needed considering the benefits of forest management for carbon sequestration against the possible impacts of this management on biodiversity and the cultural and ecological values of undisturbed and old-growth forests.

4.5   Net zero in Stockholm County

In 2015, published GHG emissions in Stockholm County were 5.18 $MtCO_2$-eq, and we calculated the sequestration from the forests in the county to be 2.36 $MtCO_2$-eq for the same year. More recently, the emissions for Stockholm County in 2022 were 3.71 $MtCO_2$-eq (Region Stockholm, 2024; Swedish Meteorological and Hydrological Institute, 2024). There is a clear decrease in emissions from 2015 to 2022, and while the 2015 sequestrations could only offset around 45% of the 2015 emissions, those same sequestrations would have offset nearly 64% of the 2022 emissions. In their climate action planning and climate budget, Region Stockholm aims for total GHG emissions of around 1 $MtCO_2$-eq/year by 2040, with further decreases continuing through 2045 and 2050 (Region Stockholm, 2022; Tillväxt- och Regionplaneförvaltningen, 2019). If such substantial emissions reductions can be realized, our calculated sequestration potential of 2.99 $MtCO_2$-eq in 2040 would be more than sufficient to offset the remaining emissions by 2040. Even if the uncertainty inherent in these calculations means that the actual sequestration is considerably lower than this in 2040, it is still very likely to be sufficient to offset the remaining GHG emissions in the county.

However, it is worth noting that the official emissions mentioned previously do not include consumption-based emissions for the residents of Stockholm County. Consumption emissions include those emissions associated with, for example, the production of food and goods consumed by people in Stockholm, as well as those associated with international travel, investments, and more. In 2014, these emissions were calculated to be 9.83 tCO_2-eq per person in Sweden, which comes to a total of 21.6 $MtCO_2$-eq when multiplied by the population of Stockholm County for that year. By 2021, per capita emissions in Sweden had decreased to 8.44 tCO_2-eq, which comes to a total of 20.4 $MtCO_2$-eq when multiplied by the 2021 population, which increased over the same period (Naturvårdsverket, 2021; Region Stockholm, 2023). In their climate action plan, Region Stockholm mentions these emissions and sets a goal of halving them by 2030 (based on 2014 levels), which would result in per capita emissions of just under 5 tCO_2-eq (Tillväxt- och Regionplaneförvaltningen, 2019). If this is achieved, the total consumption emissions in Stockholm County will be 12.7 $MtCO_2$-eq, based on population projections for 2030 (Region Stockholm, 2023).

If consumption emissions are included in net zero calculations, they would need to decrease very significantly to below 0.75 tCO_2-eq per capita by 2040 (from more than 8 tons per capita in 2021) for our projected 2040 forest sequestrations to offset these emissions in addition to the official local emissions discussed earlier. Although some measures to decrease these emissions are included in the regional CAP, achieving these reductions relies largely on changes in behavior by individuals living in Stockholm and the actions of external actors such as international manufacturing and shipping companies. This highlights again the importance of protecting the region's forests and a focus on management to increase carbon sequestration if Stockholm County is to approach true net zero emissions by 2040, even with significant climate action locally and radical changes in consumer behavior.

5 Conclusions

The forests in Stockholm County had a large sequestration potential in 2015, and this is projected to increase by more than 25% by 2040 due to forest expansion and aging, meaning that they could be a significant aid in reducing net carbon emissions. While forest aging and some impacts of climate change are likely to lead to increased sequestration capacity, these forests still need to be managed effectively if they are to be and remain effective long-term nature-based solutions for climate change mitigation. Management can not only prevent loss to land use change but also help to ensure that they remain resilient against the risks and potential negative impacts of climate change. Forestry and harvesting practices can be adapted to ensure that carbon sequestration potential is increased and maintained in the future. The sequestration offered by Stockholm County's forests could be sufficient to offset local emissions by 2040 if climate action goals are achieved. However, achieving true net zero in Stockholm when considering consumer emissions will still require very significant consumer behavior and societal change beyond what is currently planned, as these emissions are significantly larger than the sequestration potential offered by the forests in the region. Further research is needed to reduce uncertainty in the calculations of forest carbon sequestration potentials, particularly if they are to be relied upon in carbon accounting for net zero goals.

References

Bascietto, M., Cherubini, P., & Scarascia-Mugnozza, G. (2004). Tree rings from a European beech forest chronosequence are useful for detecting growth trends and carbon sequestration. *Canadian Journal of Forest Research*, 34, 481–492. https://doi.org/10.1139/x03-214.

Bradford, J. B., & Kastendick, D. N. (2010). Age-related patterns of forest complexity and carbon storage in pine and aspen–birch ecosystems of northern Minnesota, USA. *Canadian Journal of Forest Research*, 40, 401–409. https://doi.org/10.1139/X10-002.

Burke, T., Rowland, C., Whyatt, J. D., Blackburn, G. A., & Abbatt, J. (2021). Achieving national scale targets for carbon sequestration through afforestation: Geospatial assessment of feasibility and policy implications. *Environmental Science and Policy*, 124, 279–292. https://doi.org/10.1016/j.envsci.2021.06.023.

Clay, C., Nave, L. E., Nadelhoffer, K. J., Den Uyl, J., Propson, B., & Gough, C. M. (2020). *Net ecosystem production across successional time in a North American eastern temperate forest chronosequence. 2020* (pp. B064–0019).

Colding, J. (2013). Local assessment of Stockholm: Revisiting the Stockholm urban assessment. In T. Elmqvist, M. Fragkias, J. Goodness, B. Güneralp, P. J. Marcotullio, R. I. McDonald, … C. Wilkinson (Eds.), *Urbanization, biodiversity and ecosystem services: Challenges and opportunities: A global assessment* (pp. 313–335). Netherlands, Dordrecht: Springer. https://doi.org/10.1007/978-94-007-7088-1_17.

d'Annunzio, R., Sandker, M., Finegold, Y., & Min, Z. (2015). Projecting global forest area towards 2030. *Forest Ecology and Management*, *352*, 124–133. https://doi.org/10.1016/j.foreco.2015.03.014. Changes in Global Forest Resources from 1990 to 2015.

Dar, J. A., Subashree, K., Bhat, N. A., Sundarapandian, S., Xu, M., Saikia, P., Kumar, A., Kumar, A., Khare, P. K., & Khan, M. L. (2020). Role of major forest biomes in climate change mitigation: An eco-biological perspective. In N. Roy, S. Roychoudhury, S. Nautiyal, S. K. Agarwal, & S. Baksi (Eds.), *Socio-economic and eco-biological dimensions in resource use and conservation: Strategies for sustainability* (pp. 483–526). Cham: Springer International Publishing. https://doi.org/10.1007/978-3-030-32463-6_24.

Davis, E. C., Sohngen, B., & Lewis, D. J. (2022). The effect of carbon fertilization on naturally regenerated and planted US forests. *Nature Communications*, *13*, 5490. https://doi.org/10.1038/s41467-022-33196-x.

De Simon, G., Alberti, G., Delle Vedove, G., Zerbi, G., & Peressotti, A. (2012). Carbon stocks and net ecosystem production changes with time in two Italian forest chronosequences. *European Journal of Forest Research*, *131*, 1297–1311. https://doi.org/10.1007/s10342-012-0599-4.

Eriksson, O. (2020). Origin and development of managed meadows in Sweden: A review. *Rural Landscapes: Society, Environment, History*, *7*, 2. https://doi.org/10.16993/rl.51.

Felton, A., Gustafsson, L., Roberge, J.-M., Ranius, T., Hjältén, J., Rudolphi, J., Lindbladh, M., Weslien, J., Rist, L., Brunet, J., & Felton, A. M. (2016). How climate change adaptation and mitigation strategies can threaten or enhance the biodiversity of production forests: Insights from Sweden. *Biological Conservation*, *194*, 11–20. https://doi.org/10.1016/j.biocon.2015.11.030.

Global Forest Watch. (2024). *Fires in Stockholm, Sweden*. [WWW Document]. URL https://www.globalforestwatch.org. (Accessed 4 June 2024).

Gough, C. M., Vogel, C. S., Schmid, H. P., Su, H.-B., & Curtis, P. S. (2008). Multi-year convergence of biometric and meteorological estimates of forest carbon storage. *Agricultural and Forest Meteorology*, *148*, 158–170. https://doi.org/10.1016/j.agrformet.2007.08.004. Chequamegon Ecosystem-Atmosphere Study Special Issue: Ecosystem-Atmosphere Carbon and Water Cycling in the Temperate Northern Forests of the Great Lakes the Region.

Grelle, A., Hedwall, P.-O., Strömgren, M., Håkansson, C., & Bergh, J. (2023). From source to sink—Recovery of the carbon balance in young forests. *Agricultural and Forest Meteorology*, *330*, 109290. https://doi.org/10.1016/j.agrformet.2022.109290.

Guo, L. B., & Gifford, R. M. (2002). Soil carbon stocks and land use change: A meta analysis. *Global Change Biology*, *8*, 345–360. https://doi.org/10.1046/j.1354-1013.2002.00486.x.

Hadden, D., & Grelle, A. (2017). Net CO_2 emissions from a primary boreo-nemoral forest over a 10 year period. *Forest Ecology and Management*, *398*, 164–173. https://doi.org/10.1016/j.foreco.2017.05.008.

Hansen, K., & Malmaeus, M. (2016). Ecosystem services in Swedish forests. *Scandinavian Journal of Forest Research*, *31*, 626–640. https://doi.org/10.1080/02827581.2016.1164888.

Hansen, M. C., Potapov, P. V., Moore, R., Hancher, M., Turubanova, S. A., Tyukavina, A., Thau, D., Stehman, S. V., Goetz, S. J., Loveland, T. R., Kommareddy, A., Egorov, A., Chini, L., Justice, C. O., & Townshend, J. R. G. (2013). High-resolution global maps of 21st-century forest cover change. *Science*, *342*, 850–853. https://doi.org/10.1126/science.1244693.

Hosonuma, N., Herold, M., Sy, V. D., Fries, R. S. D., Brockhaus, M., Verchot, L., Angelsen, A., & Romijn, E. (2012). An assessment of deforestation and forest degradation drivers in developing countries. *Environmental Research Letters*, *7*, 044009. https://doi.org/10.1088/1748-9326/7/4/044009.

IPCC. (2023). *Climate change 2023: synthesis report. Contribution of working groups I, II and III to the sixth assessment report of the intergovernmental panel on climate change*. Geneva, Switzerland: IPCC.

Jaramillo, F., Lund, V., Stock, B., & Piemontese, L. (2021). *Slow-onset risks from climate change in Sweden in 2050*. Stockholm, Sweden: Swedish Civil Contingencies Agency (MSB).

Jörgensen, K., Granath, G., Lindahl, B. D., & Strengbom, J. (2021). Forest management to increase carbon sequestration in boreal *Pinus sylvestris* forests. *Plant and Soil*, *466*, 165–178. https://doi.org/10.1007/s11104-021-05038-0.

Karger, D. N., Kessler, M., Lehnert, M., & Jetz, W. (2021). Limited protection and ongoing loss of tropical cloud forest biodiversity and ecosystems worldwide. *Nature Ecology & Evolution*, *5*, 854–862. https://doi.org/10.1038/s41559-021-01450-y.

Kauppi, P., Stål, G., Arnesson Ceder, L., Hallberg-Sramek, I., Hoen, H., Svensson, A., Wernick, I., Högberg, P., Lundmark, T., & Nordin, A. (2022). Managing existing forests can mitigate climate change. *Forest Ecology and Management*, *513*, 120186. https://doi.org/10.1016/j.foreco.2022.120186.

Kuittinen, M., Moinel, C., & Adalgeirsdottir, K. (2016). Carbon sequestration through urban ecosystem services: A case study from Finland. *Science of The Total Environment*, *563–564*, 623–632. https://doi.org/10.1016/j.scitotenv.2016.03.168.

Lal, R. (2007). Carbon sequestration. *Philosophical Transactions of the Royal Society of London. Series B, Biological Sciences*, *363*, 815–830. https://doi.org/10.1098/rstb.2007.2185.

Lane, J., Greig, C., & Garnett, A. (2021). Uncertain storage prospects create a conundrum for carbon capture and storage ambitions. *Nature Climate Change*, *11*, 925–936. https://doi.org/10.1038/s41558-021-01175-7.

Laudon, H., Mosquera, V., Eklöf, K., Järveoja, J., Karimi, S., Krasnova, A., Peichl, M., Pinkwart, A., Tong, C. H. M., Wallin, M. B., Zannella, A., & Hasselquist, E. M. (2023). Consequences of rewetting and ditch cleaning on hydrology, water quality and greenhouse gas balance in a drained northern landscape. *Scientific Reports*, *13*, 20218. https://doi.org/10.1038/s41598-023-47528-4.

Lavalle, C., Kompil, M., & Aurambout, J.-P. (2015). *UI—Boundaries for the functional urban areas (LUISA Platform REF2014)*.

Li, T., Ren, B., Wang, D., & Liu, G. (2015). Spatial variation in the storages and age-related dynamics of forest carbon sequestration in different climate zones—Evidence from black locust plantations on the Loess Plateau of China. *PLoS One*, *10*, e0121862. https://doi.org/10.1371/journal.pone.0121862.

Liang, S., & Wang, J. (2020). Estimate of vegetation production of terrestrial ecosystem. In *Advanced remote sensing: Terrestrial information extraction and applications* (pp. 581–620). Academic Press.

Lindner, M., Eggers, J., & Hanewinkel, M. (2014). Climate change and European forests: What do we know, what are the uncertainties, and what are the implications for forest management? *Journal of Environmental Management*, *146*. https://doi.org/10.1016/j.jenvman.2014.07.030.

Lorenz, K., & Lal, R. (2010). *Carbon sequestration in forest ecosystems*. Netherlands, Dordrecht: Springer. https://doi.org/10.1007/978-90-481-3266-9.

Lovett, G. M., Cole, J. J., & Pace, M. L. (2006). Is net ecosystem production equal to ecosystem carbon accumulation? *Ecosystems*, *9*, 152–155. https://doi.org/10.1007/s10021-005-0036-3.

Luyssaert, S., Schulze, E.-D., Börner, A., Knohl, A., Hessenmöller, D., Law, B. E., Ciais, P., & Grace, J. (2008). Old-growth forests as global carbon sinks. *Nature, 455*, 213–215. https://doi.org/10.1038/nature07276.

Naturvårdsverket. (2020). *Svenska Marktäckedata* (p. 2000).

Naturvårdsverket. (2021). *Konsumtionsbaserade växthusgasutsläpp per person och år.* [WWW Document]. Naturvårdsverket. URL https://www.naturvardsverket.se/Sa-mar-miljon/Statistik-A-O/Vaxthusgaser-konsumtionsbaserade-utslapp-per-person/. (Accessed 26 March 2021).

Naturvårdsverket. (2023). *Tillväxt och avverkningar i skogen.* [WWW Document]. URL https://www.naturvardsverket.se/data-och-statistik/skog/skog-tillvaxt-och-avverkningar/. (Accessed 31 May 2024).

Net Zero Cities. (2022). *Meet the 112 mission cities paving the way to climate neutrality by 2030.* [WWW Document]. NetZeroCities. URL https://netzerocities.eu/mission-cities/. (Accessed 21 May 2024).

Ostrogović Sever, M. Z., Alberti, G., Delle Vedove, G., & Marjanović, H. (2019). Temporal evolution of carbon stocks, fluxes and carbon balance in pedunculate oak chronosequence under close-to-nature forest management. *Forests, 10*, 814. https://doi.org/10.3390/f10090814.

Page, J., Kåresdotter, E., Destouni, G., Pan, H., & Kalantari, Z. (2021). A more complete accounting of greenhouse gas emissions and sequestration in urban landscapes. *Anthropocene, 34*, 100296. https://doi.org/10.1016/j.ancene.2021.100296.

Pan, H., Page, J., Zhang, L., Cong, C., Ferreira, C., Jonsson, E., Näsström, H., Destouni, G., Deal, B., & Kalantari, Z. (2020). Understanding interactions between urban development policies and GHG emissions: A case study in Stockholm Region. *Ambio, 49*, 1313–1327. https://doi.org/10.1007/s13280-019-01290-y.

Petersson, H., Ellison, D., Appiah Mensah, A., Berndes, G., Egnell, G., Lundblad, M., Lundmark, T., Lundström, A., Stendahl, J., & Wikberg, P.-E. (2022). On the role of forests and the forest sector for climate change mitigation in Sweden. *GCB Bioenergy, 14*, 793–813. https://doi.org/10.1111/gcbb.12943.

Region Stockholm. (2022). *Koldioxidbudget för Stockholms län 2022.* [WWW Document]. Koldioxidbudget. URL https://www.regionstockholm.se/regional-utveckling/statistik-och-analys/klimat/koldioxidbudget/. (Accessed 28 February 2023).

Region Stockholm. (2023). *Befolkningsprognos.* [WWW Document]. Reg. Stockh. URL https://www.regionstockholm.se/regional-utveckling/statistik-och-analys/befolkning/befolkningsprognos/. (Accessed 6 June 2024).

Region Stockholm. (2024). *Utsläpp av växthusgaser.* [WWW Document]. Reg. Stockh. URL https://www.regionstockholm.se/regional-utveckling/statistik-och-analys/klimat/utslapp-av-vaxthusgaser/. (Accessed 5 June 2024).

SCB. (2023a). *Markanvändningen i Sverige efter region och markanvändningsklass. Vart 5:e år 2010-2020.* [WWW Document]. Statistikdatabasen. URL http://www.statistikdatabasen.scb.se/pxweb/sv/ssd/START__MI__MI0803__MI0803A/MarkanvN/. (Accessed 31 May 2024).

SCB. (2023b). *Nationalparker, naturreservat, naturvårdsområden, biotopskyddsområden, efter region. År 1998-2023.* [WWW Document]. Statistikdatabasen. URL http://www.statistikdatabasen.scb.se/pxweb/sv/ssd/START__MI__MI0603__MI0603D/SkyddadnaturN/. (Accessed 31 May 2024).

Skogsstyrelsen. (2024). *Skogsvårdslagen.* [WWW Document]. URL https://www.skogsstyrelsen.se/lag-och-tillsyn/skogsvardslagen/. (Accessed 6 June 2024).

Stockholms Stad. (2016). *Strategy for a fossil-fuel free Stockholm by 2040*. Stockholm: City Executive Office.

Stockholms Stad. (2023). *Traditionell ängsskötsel i Nackareservatet—Stockholms miljöbarometer*. [WWW Document]. URL https://miljobarometern.stockholm.se/natur/atgarder/stadens-mark-utanfor-stockholm/traditionell-angsskotsel-i-nackareservatet/. (Accessed 3 June 2024).

Stockholms Stad. (2020). *Climate Action Plan 2020–2023: For a fossil-free and climate-positive Stockholm by 2040*. https://miljobarometern.stockholm.se/content/docs/tema/klimat/Climate-Action-Plan-2020-2023.pdf. (Accessed 30 August 2024).

Subramanian, N., Bergh, J., Johansson, U., & Sallnäs, O. (2015). Adaptation of forest management regimes in Southern Sweden to increased risks associated with climate change. *Forests, 7*. https://doi.org/10.3390/f7010008.

Subramanian, N., Nilsson, U., Mossberg, M., & Bergh, J. (2019). Impacts of climate change, weather extremes and alternative strategies in managed forests. *Écoscience, 26*, 53–70. https://doi.org/10.1080/11956860.2018.1515597.

Swedish Meteorological and Hydrological Institute. (2023). *Advanced Climate Change Scenario Service | SMHI*. [WWW Document]. URL https://www.smhi.se/en/climate/future-climate/advanced-climate-change-scenario-service/met/sverige/medeltemperatur/rcp45/2071-2100/year/anom. (Accessed 3 June 2024).

Swedish Meteorological and Hydrological Institute. (2024). *Nationella emissionsdatabasen*. [WWW Document]. URL https://nationellaemissionsdatabasen.smhi.se/. (Accessed 5 June 2024).

Swedish Ministry of the Environment. (2019). *Sweden's fourth Biennial report under the UNFCCC*. Stockholm: Government Offices of Sweden.

Swedish University of Agricultural Sciences. (2015). *SLU Skogskarta*. [WWW Document]. SLU. SE. URL https://www.slu.se/centrumbildningar-och-projekt/riksskogstaxeringen/statistik-om-skog/slu-skogskarta/. (Accessed 30 May 2024).

Swedish University of Agricultural Sciences. (2017). *Skogsdata 2017*. Umea.

Swedish University of Agricultural Sciences. (2022). *Skogsdata 2022*. Umea.

Taylor, A. R., Seedre, M., Brassard, B. W., & Chen, H. Y. H. (2014). Decline in net ecosystem productivity following canopy transition to late-succession forests. *Ecosystems, 17*, 778–791. https://doi.org/10.1007/s10021-014-9759-3.

Tillväxt- och Regionplaneförvaltningen. (2017). *Regional Utvecklingsplan För Stockholmsregionen (RUFS) 2050*. [WWW Document]. URL http://www.rufs.se/publikationer/2018/rufs-2050/. (Accessed 4 February 2020).

Tillväxt- och Regionplaneförvaltningen. (2019). *Klimatfärdplan 2050 för Stockholmsregionen*. Stockholm: Region Stockholm.

UNFCC. (2022). *Nationally determined contributions registry*. [WWW Document]. URL https://unfccc.int/NDCREG. (Accessed 6 June 2024).

Uri, V., Kukumägi, M., Aosaar, J., Varik, M., Becker, H., Morozov, G., & Karoles, K. (2017). Ecosystems carbon budgets of differently aged downy birch stands growing on well-drained peatlands. *Forest Ecology and Management, 399*, 82–93. https://doi.org/10.1016/j.foreco.2017.05.023.

Uri, V., Varik, M., Aosaar, J., Kanal, A., Kukumägi, M., & Lõhmus, K. (2012). Biomass production and carbon sequestration in a fertile silver birch (*Betula pendula* Roth) forest chronosequence. *Forest Ecology and Management, 267*, 117–126. https://doi.org/10.1016/j.foreco.2011.11.033.

Varik, M., Kukumägi, M., Aosaar, J., Becker, H., Ostonen, I., Lõhmus, K., & Uri, V. (2015). Carbon budgets in fertile silver birch (*Betula pendula* Roth) chronosequence stands. *Ecological Engineering, 77,* 284–296. https://doi.org/10.1016/j.ecoleng.2015.01.041.

Yanai, R. D., Currie, W. S., & Goodale, C. L. (2003). Soil carbon dynamics after forest harvest: An ecosystem paradigm reconsidered. *Ecosystems, 6,* 197–212. https://doi.org/10.1007/s10021-002-0206-5.

Zhu, K., Song, Y., & Qin, C. (2019). Forest age improves understanding of the global carbon sink. *Proceedings of the National Academy of Sciences, 116,* 3962. https://doi.org/10.1073/pnas.1900797116.

Chapter 1.2

Impact of green, gray, and hybrid infrastructure on flood risk in partly urbanized catchment

Kristina Unger[a], Mojca Šraj[a], Jiří Jakubínský[b], and Nejc Bezak[a]
[a]*Faculty of Civil and Geodetic Engineering, University of Ljubljana, Ljubljana, Slovenia,* [b]*Global Change Research Institute CAS, Brno, Czech Republic*

1 Introduction

One of the most urgent issues facing the world today is climate change (IPCC, 2012, 2019). It is expected that the frequency and magnitude of natural phenomena caused by climate change will increase so that the population will suffer more and more from the direct and indirect consequences of these events (Raymond et al., 2020). In particular, climate change or climate variability may be one of the main causes of changes in rainfall patterns in different parts of the world and may lead to higher rainfall intensities and consequently more severe flooding (Blöschl et al., 2019; Hannaford, 2015). Recently, the frequency of floods has been increasing, causing enormous damage not only in Europe but also worldwide (Kundzewicz et al., 2010). Moreover, floods are considered the most common natural disaster in Europe (Kundzewicz et al., 2010), causing significant damage and threatening human lives every year. It is predicted that the risk of flooding will increase mainly as a result of climate change (Blöschl et al., 2017; Winsemius et al., 2016). At the same time, urbanization may exacerbate the situation (Galli et al., 2021; Rosenzweig et al., 2018).

Nowadays, many scientists and professionals from water-related disciplines around the world are striving to mitigate the problem of increasing flood risks. The higher probability of flood occurrence makes it necessary to adopt various measures to reduce the associated risks (Alves et al., 2020; Anderson et al., 2022). To adapt to highly variable weather conditions in combination with other occurring changes, such as urbanization, various gray, green, blue, and hybrid flood mitigation measures can be implemented today (Anderson et al., 2022; Hartmann et al., 2019; Kabisch et al., 2017; Kryžanowski et al., 2014;

Nature-Based Solutions in Supporting Sustainable Development Goals
https://doi.org/10.1016/B978-0-443-21782-1.00004-X

25

Vozelj et al., 2023). Due to the unpredictability and uniqueness of each flood disaster, different risk mitigation strategies and techniques are now being considered for different flood events (Huang et al., 2020; Kundzewicz et al., 2010; Nakamura, 2022). The encroachment of humans into floodplains makes the issue of protecting people from flooding even more important due to increased vulnerability (Kundzewicz et al., 2010). Therefore, it is important to find the most effective and practical solutions to mitigate the negative impact of floods. Despite the global implementation of gray infrastructure and its widespread acceptance by the public, nature-based solutions are still on their way to general acceptance, and there are still many limitations related to the implementation of nature-based solutions (Anderson et al., 2022; Raška et al., 2022; Roberts et al., 2023). Moreover, gray measures are still the predominant solution in many areas around the globe due to limited reliance on purely nature-based solutions (Anderson et al., 2022). In general, although several researches have been conducted focusing on investigating the effectiveness of different gray, green, and blue measures and their added benefits in terms of reducing the impact of flooding (Kryžanowski et al., 2014; Pudar et al., 2020; Štajdohar et al., 2016; Vozelj et al., 2023), there is still a lack of sufficient research in this area in terms of different flood risk reduction options and providing evidence-base for flood risk reduction (Roberts et al., 2023; Zabret & Šraj, 2015). Therefore, it is necessary to conduct additional research aimed at defining a set of technical- and nature-based solutions for flood protection for specific catchments. Furthermore, much of the research focuses primarily on just one measure to reduce the impact of flooding (Johnen et al., 2020); however, additional studies on the effectiveness of multiple measures and their combinations or hybrid solutions are needed to find a better and an optimal approach to tackle the problem of floods.

Therefore, in this study, the partly urbanized Glinščica River catchment in Slovenia was used to analyze the impact of selected mitigation measures on flood risk. The idea is to analyze the impact of the green, gray, and hybrid infrastructure on flood risk in the city of Ljubljana, in order to highlight the most appropriate measures for this area. Hence, we investigate the effectiveness of different gray, green, and hybrid solutions for flood risk management. The main objective of this research is to evaluate the performance of various flood risk measures (i.e., green, gray, and hybrid).

2 Data and methods

2.1 Flood protection measures

To deal with the changing occurrence and characteristics of floods (Kemter et al., 2020), different types of measures need to be implemented, including green, blue, gray, and hybrid measures (Alves et al., 2018, 2020; Kabisch et al., 2016, 2017). Although their application is currently at the forefront, purely green measures may often not be sufficient to cope with projected future climate hazards (Anderson et al., 2022; Bezak et al., 2021; Johnen et al., 2020; Kabisch et al., 2017; Vozelj et al., 2023; Zabret & Šraj, 2015). Therefore, hybrid solutions that combine elements of green and gray infrastructure or combine green and gray measures within one study area seem to be an attractive option for climate change adaptation (Alves et al., 2020; Kabisch et al., 2017; Nakamura, 2022; Štajdohar et al., 2016), as they better reflect the diversity of environmental conditions (urban centers and rural hinterlands), the uncertainties related to future hazard frequency, the opportunities arising from specific land ownership (amount of land required to implement the measure), and the social acceptance of these measures. Unger (2023) carried out a review of multiple green, gray, and hybrid flood risk measures. For each measure, a standardized review report was prepared covering aspects like feasibility, cost-effectiveness, maintenance, and climate change impact. The selected measures were classified into the following three categories: green, gray, and hybrid measures. Traditional and, to a lesser extent, conventional flood mitigation infrastructures have been selected regarding the gray measures. Gray measures visually represent rigid infrastructure usually made of hardly degradable materials, e.g., concrete or steel, which are already known to have a prevailing gray visual effect compared to other flood risk reduction techniques (Unger, 2023). In addition, such measures provide limited or almost no ecosystem services (Unger, 2023). On the contrary, green measures tend to have more favorable ecosystem functions than all other categories of flood risk mitigation and are mostly composed of biodegradable materials (Roberts et al., 2023). Even though certain technical equipment is usually required in the implementation stage for the construction of green flood protection measures (Pohl & Bezak, 2022), these measures usually only have a "green" visual effect after construction (Unger, 2023). In the case of hybrid measures, flood mitigation solutions were chosen that include the functions of both gray and green protective measures. In addition, it is important to note that the combination of measures aims for solutions that are visually greener and provide a higher level of ecosystem services than similar gray measures; however, they also contain elements of gray infrastructure that should be used to properly execute them (Unger, 2023). Unger (2023) examined more than 15 different flood risk measures. However, only a limited number of these measures were selected for rainfall-runoff modeling of the partly urbanized Glinščica River catchment, as not all measures could be implemented in the HEC-HMS hydrological model or some measures had already been investigated in the previous studies (Bezak et al., 2021; Johnen et al., 2020). Consequently, the following measures were selected for consideration in this study (Table 1):

- green roofs (hybrid measure)
- urban tree cover (green measure)
- rain gardens (green measure)
- permeable sidewalks (gray measure)

TABLE 1 Short description of the selected flood protection measures.

Measure	Type of measure	Short description
Green roof	Hybrid	Green roofs are one of the solutions that not only help to cope with the increasing flood risk but also have other, no less important, benefits, such as creating a suitable environment for biodiversity development, providing thermal comfort in buildings, reducing energy consumption and pollution, and improving the esthetic appearance of buildings. The amount of ecosystem services provided is relatively high. However, during the design and construction, specific knowledge and equipment are required in order to construct green roofs.
Urban tree cover	Green	In recent decades, especially in the 20th century, with the expansion of cities and development in urban areas, there has been a tendency to cut down trees while increasing the percentage of impermeable surfaces, which has consequently led to an increase in stormwater runoff. However, trees play an important role in the water cycle; in particular, their canopies can intercept rainwater so that some of it evaporates back into the atmosphere, and the roots of trees help rainwater to percolate deeper into the soil and improve the water holding capacity of the soil. The ecosystem services provided are very high.
Rain garden	Green	A rain garden is a small garden of shrubs, flowers, grass, and other vegetation usually planted in low-lying areas on a slope to capture stormwater runoff. Rain gardens are designed to absorb excess water from rooftops, streets, lawns, and other pathways and infiltrate into the ground. The amount of ecosystem services provided is high.
Permeable sidewalk	Gray	Permeable concrete pavements can be considered as an additional measure to reduce flood risk, as they allow slow infiltration of the retained water while reducing the additional pressure on the drainage system since permeable pavements or sidewalks enable increased infiltration of water compared to classical concrete or asphalt solutions. No ecosystem services are provided.
Tree trenches	Hybrid	Stormwater tree trenches are a series of trees connected underground by a trench system to manage excess stormwater. Stormwater tree trenches also provide a healthy environment for trees to grow sustainably in urban areas where impervious sidewalks are prevalent. The specialized subsurface system consists of a soil media, inlet and outlet pipes, and a special water

TABLE 1 Short description of the selected flood protection measures—cont'd

Measure	Type of measure	Short description
		distribution system that allows stormwater to infiltrate. The amount of ecosystem services provided is relatively high.
Cistern	Gray	A cistern is similar to infiltration trench. It is a special underground system consisting of a main shaft and some other features necessary to collect stormwater runoff. In some cases, the system allows excess water to infiltrate. No ecosystem services are provided.

Detailed descriptions including additional references can be found in Unger (2023).

- tree trenches (hybrid measure)
- cisterns (gray measure)

Note that the classification of these measures into three categories (green, gray, and hybrid) is to some extent subjective but was done according to the above-mentioned criteria adopted by Unger (2023) (i.e., visual appearance, amount of ecosystem services provided, and amount of construction work needed). It should also be noted that dry retention reservoir Brdnikova and wet retention reservoir Podutik were already studied in detail by Bezak et al. (2021).

2.2 Case study description

The Glinščica River catchment is used as a case study to analyze the effectiveness of implementing gray, green, and hybrid flood risk mitigation measures based on a synthetic rainfall event (Bezak et al., 2018; Brilly et al., 2006; Šraj et al., 2010). The area of the catchment is estimated to be about $16.9 \, km^2$ (Bezak et al., 2021). It is also worth noting that the urban drainage network has also been considered in the definition of catchment area in previous studies (Šraj et al., 2010). The location of the catchment within the urban area was determined by the drainage of stormwater with the drainage-sewage system, so the orographic watershed divide does not always coincide with the contributing area of the Glinščica River catchment. The total contributing area of the Glinščica is slightly larger and covers an area of $19.3 \, km^2$, as runoff from the area between Gunclje, the railway, and the orographic watershed divide between the Glinščica and Sava catchments, as well as part of the urban areas along the estuary of Glinščica is channeled to the area of the Glinščica catchment via the sewage storm network (Šraj et al., 2010). The catchment can be classified as having a temperate continental climate with a mean precipitation of about 1500 mm per year (Johnen et al., 2020). The catchment itself is located within the municipal boundary of Ljubljana, with the Glinščica River later joining the Gradaščica stream in the downstream part (Bezak et al., 2018) (Figs. 1–3). While the upper parts of the catchment are mainly characterized by natural areas, especially the largest part of this area is covered by forest, the lower areas are more for urban and agricultural areas. The forest covers about 50% of all the catchment area, whereas the urban and agricultural areas account for around 20% each (40% in total) (Bezak et al., 2021). The river catchment is characterized by relatively steep slopes, and the elevation ranges from 210 to 590 m above sea level. Hence, the catchment has torrential characteristics, and several minor-moderate flood events have occurred in the past decades (Bezak et al., 2021, 2018; Brilly et al., 2006). The catchment time of concentration is estimated to be around 6 h (Bezak et al., 2018).

2.3 Rainfall-runoff model

The hydrological modeling was carried out using the hydrologic modeling software HEC-HMS (HEC HMS, 2021). In order to perform a hydrological simulation and thus analyze the performance of each selected measure in terms of reducing peak discharges and outflow volumes during runoff events, previously selected measures were included in this model. The model was already set up and used in previous studies (Bezak et al., 2021; Johnen et al., 2020; Šraj et al., 2010). Prior to modeling, it was necessary to obtain average curve numbers (CNs) and lag-time parameters based on land use, soil, and catchment characteristics. As for the calibration and validation of the model, this was already carried out in the previous studies based on measured discharge data. More information is provided by Šraj et al. (2010). In this study, the comparison

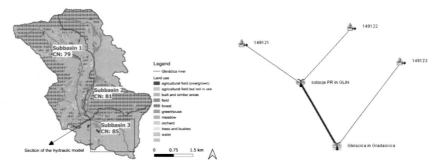

FIG. 1 The Glinščica River catchment land use (left), and subdivision of the entire catchment into three subcatchments within the HEC-HMS modeling software (right). *(The land-use map is adopted from Bezak, N., Kovačević, M., Johnen, G., Lebar, K., Zupanc, V., Vidmar, A., Rusjan, S. (2021). Exploring options for flood risk management with special focus on retention reservoirs. Sustainability, 13. doi:10.3390/su131810099.)*

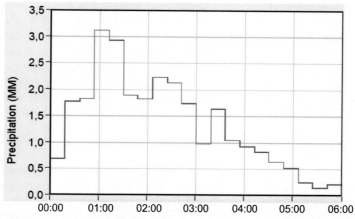

FIG. 2 Synthetic rainfall event of total duration of 6h for the return period of 25 years for the Ljubljana-Bežigrad meteorological station, used as input data for the hydrological HEC-HMS model in this study.

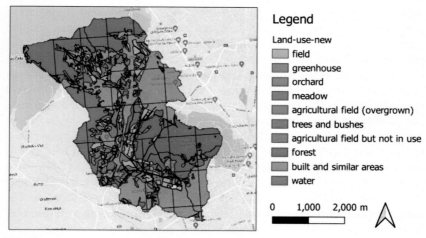

FIG. 3 Land-use map of the Glinščica River catchment.

between peak discharge values without and with the consideration of selected measures was performed using theoretical CN and lag-time parameters determined based on the land-use and soil characteristics (Unger, 2023). The SCS method using the CN parameter was applied to estimate rainfall losses (Bezak et al., 2022; HEC HMS, 2021). Furthermore, the SCS unit hydrograph method, where the lag time for each specific case was determined based on the characteristics of each individual subcatchment, was used as a method for transforming effective rainfall into runoff (Bezak et al., 2022; HEC HMS, 2021). In the first step, the model was run for the case without any measure in order to obtain results for the initial condition. In the second step, the modeling was calculated considering the application of the specific measure as described in the following sections. In terms of modeling, the entire catchment was divided into three subcatchments for model simplicity (Fig. 1). The types of land use in the Glinščica River catchment as well as their subdivision into subcatchments are shown in Fig. 1. The focus of this study is mainly on the outflow element of the Glinščica River catchment (i.e., most downstream part of the catchment).

2.4   Synthetic rainfall event

Synthetic (design) rainfall events of 2-, 10-, and 25-year return periods were used to analyze the effectiveness of the selected flood protection measures in terms of reduction of peak discharge and flow volumes in the Glinščica River catchment. The design rainfall events are known as hypothetical rainfall events of a certain duration and intensity, which can be used as input for the analysis of flood risk mitigation options. These rainfall events were derived in this study using Huff curves and intensity-duration-frequency (IDF) curves, as described in the previous works (Bezak et al., 2018; Dolšak et al., 2016). Huff curves, in turn, were originally developed by Huff (1967) and represent a method of expressing dimensionless rainfall depth-duration curves of a given precipitation station or area (Dolšak et al., 2016). These curves can help to generate input precipitation (synthetic/design rainfall events) for a hydrological model, as was done in the previous studies on the Glinščica River catchment (Bezak et al., 2018). Huff curves were developed by Dolšak et al. (2016) for 30 different rainfall stations in Slovenia based on the available historical precipitation data, one of which, Ljubljana-Bežigrad, is very close to the studied Glinščica River catchment. For this station, Dolšak et al. (2016) developed dimensionless curves for four different time periods: 3–6, 6–12, 12–24, and >24 h. In this study, the 6-h synthetic rainfall events developed by Dolšak et al. (2016) were used, as the time of concentration in the Glinščica River catchment is 6 h. Fig. 2 shows the 6-h synthetic rainfall event for the return periods of 25 years. The same temporal distribution but different total rainfall amounts were used for 2- and 10-year return periods as well (Unger, 2023).

2.5   Modeling of green roofs

This section explains the modeling procedure using green roofs as an example.

(1) In the first step, all land-use types of the three subcatchments were calculated based on the land-use map of the area (Fig. 3).
(2) As a next step, it was necessary to define an area of interest (i.e., green roofs can only be implemented in urban areas) within the "built and similar areas" in order to find a proportion of roof cover in relation to the defined area of interest (Fig. 4).

FIG. 4   Area of interest within the "built-up and similar areas" type of land use.

(3) Once the area of interest was defined, the potential green roof area was delineated based on the available roof cover (Fig. 5).

(4) The ratio between the potential green roof area and the defined area of interest was determined to be 30% (Fig. 5). This was necessary to determine the approximate area of green roofs in each subcatchment. Table 2 shows the area of the defined polygon (area of interest), the area of the green roof cover, the calculated percentage of green roof cover in the defined area of interest, and the CN parameter of the new land-use type. The CN of the green roof cover itself was obtained using the Green Values Calculator (CNT, 2024), which calculates CNs based on the defined area and land-use types.

(5) The next step was to develop possible scenarios for the green roofs. As shown in Table 3, a total of four different scenarios were defined for further

FIG. 5   Potential green roof cover within the selected area of interest used to define the percentage of green roof area.

TABLE 2 Roof cover and the CN parameter for green roofs.

Area of interest (obtained from QGIS), (m^2)	235,879
Roof cover (obtained from QGIS), (m^2)	69,730
Fraction of roof cover relative to the area of interest (%)	~30
Curve number (CN) of green roof cover (CNT, 2024)	80

TABLE 3 Different scenarios used for modeling the green roofs.

Scenarios	
Scenario 1	30% of "built and similar areas" in all three subcatchments are green roofs
Scenario 2	30% of "built and similar areas" in subcatchment 1 are green roofs
Scenario 3	30% of "built and similar areas" in subcatchment 2 are green roofs
Scenario 4	30% of "built and similar areas" in subcatchment 3 are green roofs

modeling in HEC-HMS. Scenario 1 implies the implementation of the green roofs in all three subcatchments, while scenarios 2, 3, and 4 refer to the implementation of the green roofs only in subcatchments 1, 2, and 3, respectively. Finally, as green roofs represent 30% of the total built-up area in each subcatchment, the following scenarios were established, as shown in Table 3.

(6) The next step was to calculate the average CN of each subcatchment for the baseline situation (before the green roofs were implemented) and for the case in which the green roofs were implemented. Knowing the percentage of green roofs in relation to the total built-up area, the area of each land-use type, and their corresponding CNs, the following average CNs were calculated for each subcatchment (Table 4).

(7) For each scenario, the lag-time parameter was calculated using the empirical formula that uses the maximum retention, the hydraulic length of the catchment, and the slope of the catchment (HEC HMS, 2021; Johnen et al., 2020; Unger, 2023). Based on the change in land use (with the implementation of the specific scenario), the lag time and CN parameters were derived for each scenario.

The abovementioned CN parameters and lag times were then substituted into the HEC-HMS model of the Glinščica River catchment for each scenario and return period investigated (2-, 10-, and 25-year return periods) to obtain

TABLE 4 Initial average CN (shown at the bottom of the table) without green roofs (left table) and average CN (shown at the bottom of the table) with green roofs (right side) for subcatchment 1 (units of total area: m²).

Initial state: subbasin 1			Scenario 1, 2, 3, 4: subbasin 1		
CN	**Total area**	**Land use**	**CN**	**Total area**	**Land use**
86	468,347	Field	86	468,347	Field
86	1499	Greenhouse	86	1499	Greenhouse
86	57,904	Orchard	86	57,904	Orchard
72	1,501,286	Meadow	72	1,501,286	Meadow
91	100,442	Agricultural field (overgrown)	91	100,442	Agricultural field (overgrown)
74	120,304	Trees and bushes	74	120,304	Trees and bushes
91	35,343	Agricultural field but not in use	91	35,343	Agricultural field but not in use
74	2,966,239	Forest	74	2,966,239	Forest
91	1,812,024	Built and similar areas	91	1,268,417	Built and similar areas
			80	543,607	Green roofs
99	12,212	Water	99	12,212	Water
79	**7,075,600**		**78**	**7,075,600**	

The highlighted cells indicate land-use types that were changed.

the hydrological modeling results using the synthetic rainfall events. Note that the same modeling procedure was performed also for the other five selected measures that are described in the following subsection. The initial CNs of each flood mitigation measure were determined using the stormwater management calculator (CNT, 2024) as in the case of green roofs. Additional information and details about applied modeling are provided in Unger (2023).

2.6 Modeling of other selected flood mitigation measures

Table 5 provides a summary of the proposed scenarios for the selected flood mitigation measures. In terms of urban tree cover, the potential areas (i.e., "built-up and similar areas" type of land use) for planting new trees in each subcatchment were defined in GIS software (Unger, 2023). Based on the defined areas, it was found that approximately 14%, 14%, and 6% of the "built-up and similar areas" in subcatchments 1, 2, and 3, respectively, could be used for planting the urban trees (Table 5). This was followed by the proportion of rain gardens in the defined area, which accounts for 15% of the "built-up and similar areas." In addition, scenarios 5, 6, 7, and 8 of the same flood mitigation measure also consider the potential runoff from the roofs that can hypothetically be captured by the rain gardens. In this case, it was proposed that in addition to the area of the rain gardens themselves, which is 15% of the "built-up and similar areas," 50% of the roof runoff would be directed to the rain gardens for further infiltration (Table 5). This was calculated on the basis of the roof area previously

TABLE 5 Overview of the selected scenarios.

Measure: green roofs

Scenario 1	30% of "built and similar areas" in all three subbasins are green roofs
Scenario 2	30% of "built and similar areas" in subbasin 1 are green roofs
Scenario 3	30% of "built and similar areas" in subbasin 2 are green roofs
Scenario 4	30% of "built and similar areas" in subbasin 3 are green roofs

Measure: urban tree cover

Scenario 1	14%, 14%, and 6% of "built and similar areas" in subbasins 1, 2, and 3, respectively, are new urban tree cover
Scenario 2	14% of "built and similar areas" in subbasin 1 is new urban tree cover
Scenario 3	14% of "built and similar areas" in subbasin 2 is new urban tree cover
Scenario 4	6% of "built and similar areas" in subbasin 3 is new urban tree cover

Measure: rain gardens

Scenario 1	15% of "built and similar areas" in all three subbasins are rain gardens
Scenario 2	15% of "built and similar areas" in subbasin 1 are rain gardens
Scenario 3	15% of "built and similar areas" in subbasin 2 are rain gardens
Scenario 4	15% of "built and similar areas" in subbasin 3 are rain gardens
Scenario 5	15% of rain gardens+50% of runoff from roofs in all three subbasins
Scenario 6	15% of rain gardens+50% of runoff from roofs in subbasin 1
Scenario 7	15% of rain gardens+50% of runoff from roofs in subbasin 2
Scenario 8	15% of rain gardens+50% of runoff from roofs in subbasin 3

Measure: permeable sidewalks

Scenario 1	6% of "built and similar areas" in all three subbasins are permeable sidewalks
Scenario 2	6% of "built and similar areas" in subbasin 1 are permeable sidewalks
Scenario 3	6% of "built and similar areas" in subbasin 2 are permeable sidewalks
Scenario 4	6% of "built and similar areas" in subbasin 3 are permeable sidewalks
Scenario 5	6% of permeable sidewalks+50% of runoff from roofs in all three subbasins
Scenario 6	6% of permeable sidewalks+50% of runoff from roofs in subbasin 1
Scenario 7	6% of permeable sidewalks+50% of runoff from roofs in subbasin 2
Scenario 8	6% of permeable sidewalks+50% of runoff from roofs in subbasin 3

TABLE 5 Overview of the selected scenarios—cont'd

Measure: tree trenches

Scenario 1	2% of "built and similar areas" in all three subbasins are tree trenches
Scenario 2	2% of "built and similar areas" in subbasin 1 are tree trenches
Scenario 3	2% of "built and similar areas" in subbasin 2 are tree trenches
Scenario 4	2% of "built and similar areas" in subbasin 3 are tree trenches
Scenario 5	2% of tree trenches + 50% of runoff from roads in all three subbasins
Scenario 6	2% of tree trenches + 50% of runoff from roads in subbasin 1
Scenario 7	2% of tree trenches + 50% of runoff from roads in subbasin 2
Scenario 8	2% of tree trenches + 50% of runoff from roads in subbasin 3

Measure: cisterns 1

Scenario 1	1 cistern per house in all three subbasins (in total 6560 cisterns; volume of 1 cistern $\sim$11.4 m^3)
Scenario 2	1 cistern per house in subbasin 1 (in total 2200 cisterns; volume of 1 cistern $\sim$11.4 m^3)
Scenario 3	1 cistern per house in subbasin 2 (in total 2000 cisterns; volume of 1 cistern $\sim$11.4 m^3)
Scenario 4	1 cistern per house in subbasin 3 (in total 2360 cisterns; volume of 1 cistern $\sim$11.4 m^3)

Measure: cisterns 2

Scenario 1	1 cistern per house in all three subbasins (in total 6560 cisterns; volume of 1 cistern $\sim$5.7 m^3)
Scenario 2	1 cistern per house in subbasin 1 (in total 2200 cisterns; volume of 1 cistern $\sim$5.7 m^3)
Scenario 3	1 cistern per house in subbasin 2 (in total 2000 cisterns; volume of 1 cistern $\sim$5.7 m^3)
Scenario 4	1 cistern per house in subbasin 3 (in total 2360 cisterns; volume of 1 cistern $\sim$5.7 m^3)

Please note that the land-use map is shown in Fig. 3.

identified for the green roofs. Thus, when calculating the average CNs of each subcatchment area, 50% of the roof area was also considered as the area of the rain gardens in addition to the area of the rain gardens (15% of the built-up area). For the permeable pavements, an approximate area of the pavements was calculated in the previously defined area of interest (as shown for the green roofs) by taking the average width of the pavement as 1.5 m and calculating the distance of each pavement on both sides of the road in the defined area of interest in GIS software. As a result, the pavements are found to represent almost 6% of the total area of the defined polygon (Table 5), which was then used to calculate the average CN for each subcatchment. Similar scenarios were created for the tree trenches as for the rain gardens and permeable pavements, with an additional 50% of runoff from the roads included in the last four scenarios. Thus, 50% of the road area was considered in a similar way to the rain gardens. In the case of this flood mitigation measure, it was decided to place tree trenches on both sides of the road at a distance of 3 m. Knowing the length of each road in the area, the potential number of tree trenches in the defined polygon was determined. According to CNT (2024), the area of a single tree trench is approximately 16 ft^2 ($\sim$1.5 m^2). Once the area of a tree ditch and the total number of tree ditches in the defined area of interest were known, the total area and subsequently the proportion of tree ditches in the defined area of interest could be calculated, as seen in Table 5. For infiltration shafts/dry wells, the same stormwater management calculator (CNT, 2024) was used to determine the initial CN of the measure as for green roofs and other selected flood mitigation facilities. In this case, it was not possible to calculate the proportion of the measure in relation to the total built-up area based on its surface area, as dry wells are usually underground. It was therefore necessary to find another solution similar to that used for the other measures that would not affect the final results of the study. In fact, the Rainwater Management Calculator (CNT, 2024) offers its users an option that allows them to define the initial characteristics of any subcatchment and, as a result, to obtain the average CN of the defined subcatchment. In addition, the same tool can be used to implement one of the available flood mitigation measures (in this case, dry wells) and then obtain the average CN of the same subcatchment after the implementation of the selected measure. However, the tool showed no change in the average CN of the defined subcatchment, which, as it turned out later, was due to the small volume of dry wells. To solve this problem, it was decided to replace the dry wells with cisterns, which perform more or less the same function but have a larger volume. In this case, the following two types of cisterns were selected for further analysis: 3000-gal cisterns ($\sim$11.4 m^3) and 1500-gal cisterns ($\sim$5.7 m^3) (Table 5).

3   Results and discussion

Synthetic rainfall events were used to analyze the performance of the selected flood mitigation measures in terms of flood risk reduction in the Glinščica River catchment for three selected return periods (i.e., 2, 10, and 25 years). As can be seen from Tables 6 and 7, the results of the hydrological modeling are expressed in terms of the percentage difference between two different states of the model: before and after the application of one of the selected flood mitigation measures.

TABLE 6 Example of rainfall-runoff modeling results for the return period of 2 years for all the elements of the hydrological model (shown in Fig. 1) of the Glinščica River catchment.

	2-Year return period: green roofs			
	Hydrologic element (shown in Fig. 1)	Drainage area (km^2)	Peak discharge (m^3/s)	Volume (mm)
Before applying green roofs	sotocje PR in GLIN	13.19	13.7	12.55
	Glin in Grad	16.85	17.7	13.77
	Reach-1	13.19	13.2	12.55
	149121	7.20	7.00	11.66
	149122	5.99	6.70	13.61
	149123	3.66	6.10	18.17
After applying green roofs	sotocje PR in GLIN	13.19	12.7	11.60
	Glin in Grad	16.85	16.2	12.51
	Reach-1	13.19	12.3	11.60
	149121	7.20	6.50	10.77
	149122	5.99	6.20	12.61
	149123	3.66	5.20	15.77

3.1 Green roofs

To better understand the percentage difference used to express the results obtained, green roofs are used as an example in this section. Table 6 shows the results for the runoff volume and peak discharge of different hydrological elements of the HEC-HMS model (Glinščica River catchment) developed for the green roofs (2-year return period, scenario 1). In the first step, the model was set up for the case without green roofs in order to obtain results for the initial condition in which the measure was not applied. In the second step, the modeling was calculated considering the application of the green roofs as described in the previous sections. In this case, the CNs and lag-time parameters were changed in order to observe the effectiveness of the applied flood mitigation measure in terms of peak discharge and runoff volume reduction (Table 6). Table 7 shows the percentage difference between the results of the two cases presented in Table 6. In addition, the results for the 10- and 25-year return periods of the same measure are also shown in Table 7. Here, the percentage difference (before and after application of the green roofs) in peak discharge

TABLE 7 Example of the green roofs modeling results for scenario 1 for all three return periods (i.e., 2-, 10-, and 25-year return periods).

| | Difference % (before and after applying green roofs for scenario 1) | | | | | |
| | 2-Year return period | | 10-Year return period | | 25-Year return period | |
Hydrological elements as shown in Fig. 1	Peak discharge %	Volume %	Peak discharge %	Volume %	Peak discharge %	Volume %
sotocje PR in GLIN	7.3	7.6	5.9	5.3	5.2	4.6
Glin in Grad	8.5	9.2	5.8	6.4	5.3	5.6
Reach-1	6.8	7.6	5.4	5.3	4.9	4.6
149121	7.1	7.6	6.0	5.4	5.0	4.7
149122	7.5	7.3	5.8	5.2	4.9	4.5
149123	14.8	13.2	9.6	9.6	8.8	8.4

for the 2-year return period is between 6.8% and 14.8%, depending on the hydrological element of the model (Fig. 1). It can also be observed that as the return period increases, the percentage difference decreases for both volume and peak discharge, indicating the greater effectiveness of the flood mitigation measure for the smaller magnitude events.

3.2 All selected measures

Table 8 summarizes the results for all tested scenarios for the outflow element (point). This element represents the final point of the Glinščica River catchment, and all results presented in this section relate specifically to this element. To effectively compare the performance of the selected flood mitigation measures in each specific scenario, the results of each measure are separated and distributed according to the designated scenarios, as shown in Table 8.

Rain gardens exhibited the most favorable outcomes among all tested flood mitigation strategies in terms of minimizing peak discharge and outflow volume reduction under almost all circumstances. Scenario 1 showed the greatest difference between the hydrologic conditions before and after the application of the specific measure when all flood mitigation measures in scenarios 1–4 were examined (Table 8). In this scenario, it is assumed that the flood mitigation measures were implemented in all three subcatchments of the Glinščica River, resulting in increased effectiveness in reducing flood risk. Rain gardens were found to be the most effective measure for the 2-year return period, demonstrating a reduction in peak discharge and volume by 13% and 15%, respectively. Following these results, it was determined that the flood risk mitigation capabilities of the green roofs and stormwater cisterns 1 (with a volume of approximately 11.4 m^3) were superior to the other measures. In this scenario, permeable sidewalks and cisterns 2 (with a volume of approximately 5.7 m^3) produced the lowest result (i.e., less than 1% difference) for the 25-year return period. Cisterns 2 ranked last among all measures in terms of their ability to reduce flood peak discharge and volume.

While rain gardens demonstrated their superiority over other flood mitigation options in scenario 1, scenario 2 indicated that green roofs, rain gardens, urban tree cover, and cisterns all yielded similar reductions in peak discharges. However, in scenario 2, the difference between the results before and after implementing the measures was smaller than in scenario 1, as the measures were only applied in subcatchment 1. Additionally, in this scenario, permeable pavements, stormwater storage tanks, and tree trenches had no impact on flood risk reduction and remained the least effective measures, as seen in the previous scenario (Table 8). In scenarios 3 and 4, rain gardens again demonstrated their supremacy in reducing flood risk in the selected case study compared to other measures. In scenarios 3 and 4 (rain gardens), the differences in peak discharge were between 4.0%–5.6% and 2.5%–4.0%, respectively. Similarly, the reduction in volume by the same measure showed a range between 3.2%–5.0% and

TABLE 8 Main-modeling results (% difference between before and after applying specific measures) for all considered flood protection measures for all three return periods.

Difference % (before and after applying the measure)	Synthetic rainfall events: scenarios 1–4					
	2-Year return period		10-Year return period		25-Year return period	
	Peak discharge %	Volume %	Peak discharge %	Volume %	Peak discharge %	Volume %
Scenario 1: measure is implemented in all subbasins						
Green roofs	8.5	9.2	5.8	6.4	5.3	5.6
Tree trenches	4	4.5	2.8	3.1	2.5	2.7
Rain gardens	13	15	9.4	10.6	8.2	9.2
Permeable sidewalks	2.8	2.6	2.2	1.8	1.9	1.6
Urban tree cover	7.3	7.3	5.2	5.2	4.8	4.5
Cisterns 1	8.5	9.2	5.8	6.4	5.3	5.6
Cisterns 2	1.1	2	0.6	1.3	0.6	1.1
Scenario 2: measure is implemented in subbasin 1						
Green roofs	3.4	2.8	2.5	2	2.3	1.8
Tree trenches	0	0	0	0	0	0
Rain gardens	3.4	2.8	2.5	2	2.3	1.8
Permeable sidewalks	0	0	0	0	0	0
Urban tree cover	3.4	2.8	2.5	2	2.3	1.8
Cisterns 1	3.4	2.8	2.5	2	2.3	1.8
Cisterns 2	0	0	0	0	0	0
Scenario 3: measure is implemented in subbasin 2						
Green roofs	2.8	2.6	2.2	1.8	1.9	1.6
Tree trenches	2.8	2.6	2.2	1.8	1.9	1.6
Rain gardens	5.6	5	4.1	3.6	4	3.2
Permeable sidewalks	2.8	2.6	2.2	1.8	1.9	1.6
Urban tree cover	2.8	2.6	2.2	1.8	1.9	1.6
Cisterns 1	2.8	2.6	2.2	1.8	1.9	1.6
Cisterns 2	0	0	0	0	0	0
Scenario 4: measure is implemented in subbasin 3						
Green roofs	2.3	3.8	1.4	2.5	1.1	2.2
Tree trenches	1.1	2	0.6	1.3	0.6	1.1
Rain gardens	4	7.2	2.8	4.9	2.5	4.2
Permeable sidewalks	0	0	0	0	0	0
Urban tree cover	1.1	2	0.6	1.3	0.6	1.1
Cisterns 1	2.3	3.8	1.4	2.5	1.1	2.2
Cisterns 2	1.1	2	0.6	1.3	0.6	1.1
Difference % (before and after applying the measure)	Synthetic rainfall events: scenarios 5–8					
	2-Year return period		10-Year return period		25-Year return period	
	Peak discharge %	Volume %	Peak discharge %	Volume %	Peak discharge %	Volume %
Scenario 5: measure is implemented in all subbasins						
Tree trenches	22	23.7	16.3	17.1	14.7	15
Rain gardens	29.9	32.6	22.3	23.8	20	20.9
Permeable sidewalks	11.3	11.6	8	8.2	7.1	7.2
Scenario 6: measure is implemented in subbasin 1						
Tree trenches	6.8	5.4	5	4.1	4.2	3.6
Rain gardens	9.6	7.9	7.4	6	6.7	5.3
Permeable sidewalks	3.4	2.8	2.5	2	2.3	1.8
Scenario 7: measure is implemented in subbasin 2						
Tree trenches	11.3	9.5	8.5	7	7.8	6.2
Rain gardens	14.1	11.6	10.5	8.6	9.7	7.6
Permeable sidewalks	5.6	5	4.1	3.6	4	3.2
Scenario 8: measure is implemented in subbasin 3						
Tree trenches	5.1	8.8	3.6	6	3.2	5.2
Rain gardens	8.5	13	5.8	9.2	4.8	8
Permeable sidewalks	2.3	3.8	1.4	2.5	1.1	2.2

Yellow color indicates the measure that has the highest reduction in the peak discharge among the tested measures. Results for the outflow location of the catchment are shown.

4.2%–7.2% for these scenarios. In scenario 3, all other flood mitigation measures performed similarly, with the exception of cisterns 2, which showed no difference compared to scenario 2 (Table 8). In scenario 4, permeable pavements, tree trenches, urban tree canopies, and cisterns 2 were found to be the less effective measures compared to other alternatives (Table 8). Scenarios 5–8 (Table 8) were established to explore three measures that account for an additional 50% runoff possibility from either roofs (via rain gardens and permeable sidewalks) or roads (via tree trenches). This runoff could hypothetically be directed into the infiltration area of flood mitigation facilities. GIS was used to determine the area of the roofs and roads, on the basis of which the aforesaid additional runoff was calculated. To increase accuracy, the area of the measure was calculated by adding 50% of the road or roof area, depending on the type of measure (tree trenches, rain gardens, or permeable sidewalks). This process was performed to determine the average CNs of each subcatchment, as explained in the "Data and methods" section. As shown in Table 8, rain gardens were the most effective option in scenarios 5–8 compared to the alternatives. Rain gardens proved most effective at reducing peak discharge (up to 9%) and volume, while permeable sidewalks were the least effective (almost zero effect on the runoff) among the three measures examined in all four scenarios (Table 8). Scenario 5, in which all three subcatchments utilize the selected flood mitigation measures, resulted in the highest percentage difference, as seen in scenario 1 (Table 8). When comparing scenarios 1 and 5, it becomes evident that in scenario 5, the efficiency of the tree trenches and permeable sidewalks (reduction in peak discharge was between 7% and 11% depending on the return period) has increased approximately fivefold compared to that of scenario 1 (Table 8). However, the effectiveness of the rain gardens (reduction in peak discharge was between 20% and 30% depending on the return period) was more than two times higher than that of scenario 1. In scenario 1, for example, the difference in both peak discharge and volume for the 2-year return period was 13% and 15% for rain gardens, respectively. In contrast, these figures rose to 29.9% and 32.6%, respectively, in scenario 5. In scenarios 5–8, scenario 7 demonstrated the highest effectiveness of the selected measures. Although the performance of the measures was about half as high as in scenario 5, all measures generally produced positive outcomes compared to the other flood mitigation facilities' scenarios.

Fig. 6 shows the discharge variance at the outflow section of the Glinščica River catchment for the 10-year return periods (scenario 1). The figure was produced using synthetic rainfall events and can visually depict the reduction in peak discharge after implementing a specific flood mitigation measure at the outlet of the Glinščica River catchments. The graph illustrates that before the measures were implemented, the initial condition highlighted in black displayed the highest peak discharge, whereas the rain gardens achieved the lowest peak discharge (Fig. 6). As demonstrated in Fig. 6, a negligible reduction in discharge was observed in the case of cisterns 2, which was previously noted. Note

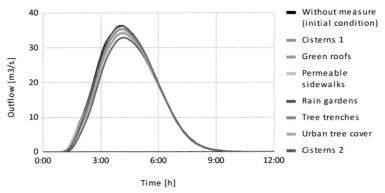

FIG. 6   Example of the simulated runoff hydrographs for the return period of 10 years for different flood mitigation measures.

that in Table 8 the results are given as a percentage difference, whereas the graph shows the computed discharge values in cubic meters per second.

3.3   Comparison of results with previous studies

It is crucial to consider other studies that have investigated comparable or alternative methods of flood mitigation and their impact on flood risk. This will enhance our understanding of the results of our research and its correlation with findings from other studies.

3.3.1   Glinščica River catchment

In the same case study area, Bezak et al. (2021) examined the implementation of various flood mitigation measures. The study investigated the effectiveness of three distinct flood mitigation methods in the Glinščica River catchment, with a specific focus on retention reservoirs. The following techniques were evaluated in the study: afforestation, permeable concrete, and dry and wet reservoirs. The research findings suggest that afforestation and permeable concrete require significantly larger areas to achieve a noteworthy decrease in peak discharge or a shift in the peak discharge timing, ultimately mitigating flood risks. Implementing these flood mitigation strategies in relatively small areas is not sufficient to achieve the desired results and substantially reduce the flood consequences. Furthermore, Bezak et al. (2021) noted that expanding these measures would result in a considerable cost escalation, thus presenting an undesirable aspect when identifying the optimal flood mitigation approach for a specific case study. However, regarding permeable concrete pavements, this measure can be considered as an additional option at a local level when a contribution to the reduction in peak discharge is required and during the reconstruction of an urbanized area. In relation to the latest flood mitigation measure examined by Bezak et al. (2021), dry and wet reservoirs can be considered as a conventional flood prevention option that can also be used for other functions like irrigation. Bezak et al. (2021) found that, although these types of reservoirs are typically considered a conventional approach to flood risk mitigation, they are more effective than afforestation and permeable pavements in terms of peak discharge reduction. The study predicts that afforestation in all three subcatchments of the Glinščica River would result in an 8%, 10%, and 10% decrease in peak discharge for the 25-, 10-, and 2-year return periods, respectively. For the two reservoirs Podutik and Brdnikova, located in the Glinščica River catchment, the reduction in peak discharge for the same return periods was calculated to be 33%, 43%, and 46%, respectively.

3.3.2   Other locations

Additionally, a study beyond the scope of this research (outside the Glinščica River catchment) was conducted by Te Linde et al. (2010), investigating the effectiveness of various flood management measures in the Rhine River catchment to reduce peak discharges under an extreme climate change scenario for the year 2050. The measures analyzed encompass retention polders, afforestation, meandering of the Upper Rhine River, reforestation of floodplains, dike heightening, and others. Te Linde et al. (2010) reported that due to climate change, the maximum water level is expected to rise by an average of 50 cm in 2050. This rise can vary from a few centimeters up to 137 cm. The study showed that the current flood mitigation options in the Rhine River catchment, as well as those planned by the Rhine Action Plan on Floods, are not sufficient to cope with the future increase in flood magnitudes and probabilities resulting from climate change. However, Te Linde et al. (2010) concluded that raising the dike by 1.29–3.25 m is the only option to mitigate flooding and reduce risks at specific sites in the studied case. Another study by Dietz and Clausen (2005) evaluated the effectiveness of rain gardens in controlling runoff and improving water quality. As in the previous research, Dietz and Clausen (2005) found that rain gardens have a high infiltration capacity and can be considered an effective method of reducing peak flow rates, as only 0.8% of the runoff in the study was uncontrolled.

In the research conducted by Liu et al. (2014), the effectiveness of green infrastructure in reducing urban flood runoff and peak flow in Beijing was investigated. The study evaluated flood mitigation measures, including the expansion of green areas, the introduction of permeable pavements, and the use of runoff retention facilities. The results indicated that pervious pavements have a greater potential to reduce flood risk than green area expansion. Using a hydrological model designed to assess the effectiveness of a specific flood mitigation facility, Liu et al. (2014) found that replacing impermeable pavements with permeable ones can substantially alleviate flood damage in Beijing. If half of the impermeable pavements in urban areas were replaced with pervious surfaces, the peak discharge and runoff of flood events would be reduced by 37.9%–35.7% and 46.2%–42.0%, respectively, depending on the return period. Furthermore, if the area of permeable surfaces were increased from 50% to 80% by implementing pervious pavements instead of impervious ones (except for roofs), both parameters could be reduced by 54.2%–51.0% and 66.5%–59.6%, respectively. However, like the current study, Liu et al. (2014) also discovered that the capacity of a single green infrastructure to reduce runoff and peak flow is limited, especially during more severe storms. They concluded that an integrated approach is needed. The research conducted by Liu et al. (2014) suggests that the use of permeable pavements results in a greater reduction of peak discharge and runoff volume compared to the current study, which also assessed the measure's effectiveness. Although current research suggests that pervious pavements are among the least effective measures for flood mitigation, Liu et al. (2014) found relatively significant results with pervious surfaces. However, it is important to note that Liu et al. (2014) assumed extreme scenarios that may not be feasible in reality. As previously stated, Liu et al. (2014) evaluated the conversion of 50% of impervious pavements to permeable brick surfaces, as well as the full adoption of permeable pavements (excluding roofs). In contrast, our study only considers the potential replacement of pervious surfaces on urban sidewalks. When assessing the feasibility of replacing impervious urban pavements, this study considers certain aspects that were not included in Liu et al.'s (2014) study. It would be nearly infeasible to entirely substitute all current impermeable surfaces in an urban environment with permeable surfaces. Similar to permeable pavements, the study also considered the feasibility and practicability of other selected flood mitigation measures, such as tree trenches and rain gardens. To illustrate, for rain gardens, rather than selecting the entire green area, which would prove infeasible, potential measure areas were manually identified using the GIS tool in almost every house where free green spaces were available. Thus, the hydrological modeling results in this study present a relatively realistic depiction of potential flood risk reduction based on the specified features of each flood mitigation approach and subcatchment.

Roehr and Kong (2010) conducted a study to evaluate the implementation of green roofs and their impact on runoff reduction in Vancouver, Kelowna, and Shanghai. The results indicated that the deployment of green roofs could serve as a suitable measure to mitigate floods in Vancouver and Shanghai, where annual precipitation surpasses 1200 mm. The study identified that in Vancouver, the runoff reduction was between 29% and 58%, whereas in Shanghai, it was observed in the range of 28%–55%. However, the study suggests that the effectiveness of green roofs is highly dependent on the selection of appropriate plants. Moreover, the amount of rainfall during the summer period affects the efficiency of these systems. In Vancouver, low-water-requiring plants proved to be a successful option for green roofs, whereas in the second city, both high- and low-water-usage vegetation were a good choice. It is important to consider these factors when designing green roofs for optimal performance. According to Roehr and Kong (2010), rain gardens and bioswales would be a more feasible solution to reduce runoff in Kelowna due to the low density of the urban area and relatively low annual precipitation (400 mm) compared to other cities analyzed. Green roofs are less suitable for this purpose.

3.4 Study limitations and further work

This study also has some limitations that should be mentioned. For the study, it was necessary to determine the average CN of the individual subcatchments (based on the detailed land-use map of the area), as well as the CN of the new land-use types (e.g., rain gardens and green roofs). Therefore, while calculating the average CNs of the individual subcatchments, the area of all flood mitigation measures (excluding cisterns 1 and 2) was estimated manually using GIS software. This estimation method has the potential to alter the final outcomes of the study. It should be considered that a smaller area of the flood mitigation facilities within the defined polygon may alter the results of the hydrological modeling. Thus, it is crucial to consider that the final result of each approach depends on the designated area of the flood mitigation option and, as in the case of cisterns, on the volume of the structure. Additionally, the selected measures were modeled in a conceptual way without detailed representation (e.g., type of vegetation for green roofs or water movement within the measure), and detailed modeling could provide different results.

It is assumed that a combination of the most effective methods will produce more significant results in reducing flood risk. Hence, combining different measures will probably provide more efficient results. However, at this stage, this can only be considered as a suggestion for further research in the same case study.

15 Conclusions

Recently, the frequency and magnitude of floods have noticeably increased, resulting in more severe consequences and having an increasingly negative impact on the environment and the general quality of life. The aim of this study is to identify the most effective flood mitigation measures that can be implemented to reduce the risk of flooding. Different types of measures, including green (nature-based solutions), hybrid, and gray, were tested and compared. The study focuses specifically on the Glinščica River catchment as a case study. Two promising and effective measures from each category (i.e., green, gray, and hybrid) were selected for further modeling in HEC-HMS software. Specific flood mitigation measures were identified in the modeling process, including permeable sidewalks and dry wells/cisterns (gray measures), urban trees and rain gardens (green measures), and green roofs and stormwater tree trenches (hybrid measures).

The study revealed that rain gardens, classified into the "green" category, are the most effective (i.e., reduction in the peak discharge up to 30% and 13% reduction for scenario 1 for a 2-year return period) measure for mitigating floods, which could potentially be implemented in the Glinščica River catchment to reduce flood risk (among the options tested) (Table 8). After the rain gardens, green roofs (around 9% peak discharge reduction for scenario 1 for the 2-year return period) and stormwater cisterns type 1 (around 9% peak discharge reduction for scenario 1 for the 2-year return period) showed the second-highest results in terms of reducing peak discharge and volume at the outflow point of the Glinščica River catchment (Table 8). The former measures were classified as "hybrid" flood mitigation, while the latter fell under the "gray" category. According to the hydrological modeling results, cisterns type 2 and permeable sidewalks ranked last (peak discharge reduction of only 1%–3% for scenario 1 for the 2-year return period) among all flood mitigation facilities (depending on the scenario) (Table 8). However, the peak discharge reductions reported in this study were much lower compared to the values obtained using retention reservoirs.

Moreover, scenario 1 (i.e., where measures were implemented in all subcatchments) achieved the greatest reduction in peak discharge and outflow volume compared to the initial condition in all eight scenarios, with measures implemented in all three subcatchments. The study discovered that for synthetic rainfall events, the measures implemented in the flood mitigation facilities and scenarios (1–8) produced the greatest percentage difference before and after application, specifically for the 2-year return period. The impact of measures decreased with increasing return period (Table 8). This implies that the measures investigated in the study can provide more effective mitigation for smaller flood events than for more extreme flood events.

Acknowledgments

The results of this study are part of the project "Evaluation of hazard-mitigating hybrid infrastructure under climate change scenarios" funded by the Czech Science Foundation and Slovenian Research and Innovation Agency (ARIS, J6-4628). The study was partially funded also by the European Union's Horizon Europe research and innovation program under Grant Agreement 101112738.

References

Alves, A., Gersonius, B., Sanchez, A., Vojinovic, Z., & Kapelan, Z. (2018). Multi-criteria approach for selection of green and grey infrastructure to reduce flood risk and increase CO-benefits. *Water Resources Management, 32*, 2505–2522. https://doi.org/10.1007/s11269-018-1943-3.

Alves, A., Vojinovic, Z., Kapelan, Z., Sanchez, A., & Gersonius, B. (2020). Exploring trade-offs among the multiple benefits of green-blue-grey infrastructure for urban flood mitigation. *Science of the Total Environment, 703*. https://doi.org/10.1016/j.scitotenv.2019.134980.

Anderson, C. C., Renaud, F. G., Hanscomb, S., & Gonzalez-Ollauri, A. (2022). Green, hybrid, or grey disaster risk reduction measures: What shapes public preferences for nature-based solutions? *Journal of Environmental Management*, *310*, 114727. https://doi.org/10.1016/j. jenvman.2022.114727.

Bezak, N., Kovačević, M., Johnen, G., Lebar, K., Zupanc, V., Vidmar, A., & Rusjan, S. (2021). Exploring options for flood risk management with special focus on retention reservoirs. *Sustainability*, *13*. https://doi.org/10.3390/su131810099.

Bezak, N., Peranić, J., Mikoš, M., & Arbanas, Ž. (2022). Evaluation of hydrological rainfall loss methods using small-scale physical landslide model. *Water*, *14*. https://doi.org/10.3390/ w14172726.

Bezak, N., Šraj, M., Rusjan, S., & Mikoš, M. (2018). Impact of the rainfall duration and temporal rainfall distribution defined using the Huff curves on the hydraulic flood modelling results. *Geosciences*, *8*. https://doi.org/10.3390/geosciences8020069.

Blöschl, G., Hall, J., Parajka, J., Perdigão, R. A. P., Merz, B., Arheimer, B., Aronica, G. T. T., Bilibashi, A., Bonacci, O., Borga, M., Zaimi, K., Živković, N., Bloschl, G., Hall, J., Parajka, J., Perdigao, R. A. P., Merz, B., Arheimer, B., Aronica, G. T. T., … Zivkovic, N. (2017). Changing climate shifts timing of European floods. *Science (80-.)*, *357*, 588–590. https://doi.org/ 10.1126/science.aan2506.

Blöschl, G., Hall, J., Viglione, A., Perdigão, R. A. P., Parajka, J., Merz, B., Lun, D., Arheimer, B., Aronica, G. T., Bilibashi, A., Zaimi, K., & Živković, N. (2019). Changing climate both increases and decreases European river floods. *Nature*, *573*, 108–111. https://doi.org/ 10.1038/s41586-019-1495-6.

Brilly, M., Rusjan, S., & Vidmar, A. (2006). Monitoring the impact of urbanisation on the Glinscica stream. *Physics and Chemistry of the Earth*, *31*, 1089–1096. https://doi.org/10.1016/j. pce.2006.07.005.

CNT. (2024). *CNT calculator [WWW document]*. Online Calc. Retrieved from https://greenvalues. cnt.org/index.php#tabtop.

Dietz, M. E., & Clausen, J. C. (2005). A field evaluation of rain garden flow and pollutant treatment. *Water, Air, and Soil Pollution*, *167*, 123–138. https://doi.org/10.1007/s11270-005-8266-8.

Dolšak, D., Bezak, N., Šraj, M., Dolsak, D., Bezak, N., & Sraj, M. (2016). Temporal characteristics of rainfall events under three climate types in Slovenia. *Journal of Hydrology*, *541*, 1395–1405. https://doi.org/10.1016/j.jhydrol.2016.08.047.

Galli, A., Peruzzi, C., Beltrame, L., Cislaghi, A., & Masseroni, D. (2021). Evaluating the infiltration capacity of degraded vs. rehabilitated urban greenspaces: Lessons learnt from a real-world Italian case study. *Science of the Total Environment*, *787*. https://doi.org/10.1016/j. scitotenv.2021.147612.

Hannaford, J. (2015). Climate-driven changes in UK river flows: A review of the evidence. *Progress in Physical Geography*, *39*, 29–48. https://doi.org/10.1177/0309133314536755.

Hartmann, T., Slavíková, L., & McCarthy, S. (2019). *Nature-based flood risk management on private land: Disciplinary perspectives on a multidisciplinary challenge*. Springer. https://doi.org/ 10.1007/978-3-030-23842-1.

HEC HMS. (2021). *HEC HMS user's manual, v. 4.7*.

Huang, Y., Tian, Z., Ke, Q., Liu, J., Irannezhad, M., Fan, D., Hou, M., & Sun, L. (2020). Nature-based solutions for urban pluvial flood risk management. *WIREs Water*, *7*, e1421. https://doi. org/10.1002/wat2.1421.

Huff, F. A. (1967). Time distribution of rainfall in heavy storms. *Water Resources Research*, *3*, 1007. https://doi.org/10.1029/WR003i004p01007.

IPCC. (2012). *Managing the risks of extreme events and disasters to advance climate change adaptation*. Cambridge University Press.

IPCC. (2019). *Climate change and land*.

Johnen, G., Sapač, K., Rusjan, S., Zupanc, V., Vidmar, A., & Bezak, N. (2020). Modelling and evaluation of the effect of afforestation on the runoff generation within the Glinščica River catchment (Central Slovenia). In *The handbook of environmental chemistry* (pp. 1–17). Springer. https://doi.org/10.1007/698_2020_649.

Kabisch, N., Frantzeskaki, N., Pauleit, S., Naumann, S., Davis, M., Artmann, M., Haase, D., Knapp, S., Korn, H., Stadler, J., Zaunberger, K., & Bonn, A. (2016). Nature-based solutions to climate change mitigation and adaptation in urban areas: Perspectives on indicators, knowledge gaps, barriers, and opportunities for action. *Ecology and Society, 21*. https://doi.org/10.5751/ES-08373-210239.

Kabisch, N., Korn, H., Stadler, J., & Bonn, A. (2017). *Nature-based solutions to climate change adaptation in urban areas: Linkages between science, policy and practice* (1st ed.). Cham: Springer. https://doi.org/10.1007/978-3-319-56091-5.

Kemter, M., Merz, B., Marwan, N., Vorogushyn, S., & Blöschl, G. (2020). Joint trends in flood magnitudes and spatial extents across Europe. *Geophysical Research Letters, 47*. https://doi.org/10.1029/2020GL087464.

Kryžanowski, A., Brilly, M., Rusjan, S., & Schnabl, S. (2014). Review article: Structural flood-protection measures referring to several European case studies. *Natural Hazards and Earth System Sciences, 14*, 135–142. https://doi.org/10.5194/nhess-14-135-2014.

Kundzewicz, Z. W., Lugeri, N., Dankers, R., Hirabayashi, Y., Döll, P., Pińskwar, I., Dysarz, T., Hochrainer, S., & Matczak, P. (2010). Assessing river flood risk and adaptation in Europe—Review of projections for the future. *Mitigation and Adaptation Strategies for Global Change, 15*, 641–656. https://doi.org/10.1007/s11027-010-9213-6.

Liu, W., Chen, W., & Peng, C. (2014). Assessing the effectiveness of green infrastructures on urban flooding reduction: A community scale study. *Ecological Modelling, 291*, 6–14. https://doi.org/10.1016/j.ecolmodel.2014.07.012.

Nakamura, F. (2022). Concept and application of green and hybrid infrastructure. In F. Nakamura (Ed.), *Green infrastructure and climate change adaptation: Function, implementation and governance* (pp. 11–30). Singapore: Springer Singapore. https://doi.org/10.1007/978-981-16-6791-6_2.

Pohl, R., & Bezak, N. (2022). *Chapter 5: Technical and hydrological effects across scales and thresholds of polders, dams and levees*. Cheltenham, UK: Edward Elgar Publishing. https://doi.org/10.4337/9781800379534.00013.

Pudar, R., Plavšić, J., & Todorović, A. (2020). Evaluation of green and grey flood mitigation measures in rural watersheds. *Applied Sciences, 10*. https://doi.org/10.3390/app10196913.

Raška, P., Bezak, N., Ferreira, C. S. S., Kalantari, Z., Banasik, K., Bertola, M., Bourke, M., Cerdà, A., Davids, P., Madruga de Brito, M., Evans, R., Finger, D. C., Halbac-Cotoara-Zamfir, R., Housh, M., Hysa, A., Jakubínský, J., Solomun, M. K., Kaufmann, M., Keesstra, S., … Hartmann, T. (2022). Identifying barriers for nature-based solutions in flood risk management: An interdisciplinary overview using expert community approach. *Journal of Environmental Management, 310*, 114725. https://doi.org/10.1016/j.jenvman.2022.114725.

Raymond, C., Horton, R. M., Zscheischler, J., Martius, O., AghaKouchak, A., Balch, J., Bowen, S. G., Camargo, S. J., Hess, J., Kornhuber, K., Wahl, T., & White, K. (2020). Understanding and managing connected extreme events. *Nature Climate Change, 10*, 611–621. https://doi.org/10.1038/s41558-020-0790-4.

Roberts, M. T., Geris, J., Hallett, P. D., & Wilkinson, M. E. (2023). Mitigating floods and attenuating surface runoff with temporary storage areas in headwaters. *WIREs Water*, *10*, e1634. https://doi.org/10.1002/wat2.1634.

Roehr, D., & Kong, Y. (2010). Runoff reduction effects of green roofs in Vancouver, BC, Kelowna, BC, and Shanghai, P.R. China. *Canadian Water Resources Journal/Revue Canadienne des Ressources Hydriques*, *35*, 53–68. https://doi.org/10.4296/cwrj3501053.

Rosenzweig, B. R., McPhillips, L., Chang, H., Cheng, C., Welty, C., Matsler, M., Iwaniec, D., & Davidson, C. I. (2018). Pluvial flood risk and opportunities for resilience. *WIREs Water*, *5*, e1302. https://doi.org/10.1002/wat2.1302.

Šraj, M., Dirnbek, L., & Brilly, M. (2010). The influence of effective rainfall on modeled runoff hydrograph [Vplyv efekt´vnych zrážok na modelovaný hydrograf odtoku]. *Journal Of Hydrology And Hydromechanics*, *58*, 3–14. https://doi.org/10.2478/v10098-010-0001-5.

Štajdohar, M., Brilly, M., & Šraj, M. (2016). The influence of sustainable measures on runoff hydrograph from an urbanized drainage area. *Acta Hydrotechnica*, *29*, 145–162.

Te Linde, A. H., Aerts, J. C. J. H., & Kwadijk, J. C. J. (2010). Effectiveness of flood management measures on peak discharges in the Rhine basin under climate change. *Journal of Flood Risk Management*, *3*, 248–269. https://doi.org/10.1111/j.1753-318X.2010.01076.x.

Unger, K. (2023). *Evaluation of hazard-mitigating hybrid infrastructure under climate change scenarios*. University of Ljubljana.

Vozelj, U., Šraj, M., & Bezak, N. (2023). Analysis of the impact of green infrastructure on surface runoff from urban areas. *Acta Hydrotechnica*, *36*, 111–121. https://doi.org/10.15292/acta.hydro.2023.07.

Winsemius, H. C., Aerts, J. C. J. H., Van Beek, L. P. H., Bierkens, M. F. P., Bouwman, A., Jongman, B., Kwadijk, J. C. J., Ligtvoet, W., Lucas, P. L., Van Vuuren, D. P., Van Vuuren, D. P., & Ward, P. J. (2016). Global drivers of future river flood risk. *Nature Climate Change*, *6*, 381–385. https://doi.org/10.1038/nclimate2893.

Zabret, K., & Šraj, M. (2015). Can urban trees reduce the impact of climate change on storm runoff? *Urbani Izziv*, *26*, S165–S178. https://doi.org/10.5379/urbani-izziv-en-2015-26-supplement-011.

Chapter 1.3

Flood mitigation at catchment scale: Assessing the effectiveness of constructed wetlands

Elin Ekström[a], Luigia Brandimarte[a], and Carla Sofia Santos Ferreira[b,c]

[a]*Department of Sustainable Development, Environmental Science and Engineering, Sustainability Assessment and Management, KTH Royal Institute of Technology, Stockholm, Sweden,* [b]*Polytechnic Institute of Coimbra, Applied Research Institute, Coimbra, Portugal,* [c]*Research Centre for Natural Resources Environment and Society (CERNAS), Polytechnic Institute of Coimbra, Coimbra, Portugal*

1 Introduction

Past global trends have shown an increase in the frequency of floods and flood-related disasters (Adikari & Yoshitani, 2009), a trend that is expected to continue in the future, due to climate change impacts (Arnell & Gosling, 2016) and increasing urbanization (Feng et al., 2021). Traditional responses to flood risk reduction have typically involved "gray" infrastructure such as tunnels, pipes, and dikes (Vojinovic et al., 2021). However, over the past years, there has been growing recognition that working with solutions that are inspired and supported by nature has the potential to be more cost-effective and sustainable than traditional measures. This relatively new response is often referred to as nature-based solutions (NbS) (European Commission et al., 2020).

One natural system that plays a significant role in the hydrological cycle is wetlands. Wetlands provide a range of ecosystem services, the perhaps most commonly cited of which is flood reduction (Ferreira et al., 2022; Kadykalo & Findlay, 2016; Stengård et al., 2020). Wetlands are often said to "act like a sponge" by storing water during wet periods and releasing it during dry periods (Acreman & Holden, 2013; Ferreira et al., 2023). However, the understanding of wetland functions has mostly been limited to individual wetland scales, and catchment-scale wetland services have rarely been investigated (Lee et al., 2018). In fact, Quin and Destouni (2018) and Thorslund et al. (2017) both highlight the need for large-scale studies of entire wetlandscapes (i.e., multiple

Nature-Based Solutions in Supporting Sustainable Development Goals
https://doi.org/10.1016/B978-0-443-21782-1.00005-1

wetlands within a catchment). While some studies have started to take on this large-scale approach (Åhlén et al., 2020; Lee et al., 2018; Quin & Destouni, 2018), most of them focus on either natural or restored wetlands. Wetlands can also be divided into a third group, constructed wetlands, which are intentionally designed and built on a "nonwetland site," and they typically have one primary purpose such as nutrient or water retention (SMHI, 2022). Constructed wetlands become especially important when considering wetlands as NbS.

Despite their perceived value, there has been a rapid decline in wetlands worldwide due to economic drivers and human population growth (Davidson, 2014; Kreplin et al., 2021). In Sweden, almost a quarter of the country's wetland area has disappeared in the last century (Naturvårdsverket, 2021). The associated loss of ecosystem services has driven the implementation of policies, legislation, and environmental objectives that lay the foundation for the integration of wetlands and other NbS. Yet, Swedish authorities share the same challenges as the ones acknowledged by the research community. The Swedish Environmental Protection Agency (EPA) recognizes that there is a lack of knowledge and systematic evaluation of NbS, especially regarding large-scale systems and their function over time (Naturvårdsverket, 2021). This might facilitate the implementation of traditional, well-established gray solutions instead of NbS (Ferreira et al., 2023). For successful implementation, the function of NbS must be evaluated against a set of predefined goals. This, in turn, requires a robust evaluation method that enables the assessment of performance indicators (Kumar, Debele, Sahani, Rawat, Marti-Cardona, Alfieri, Basu, Basu, Bowyer, Charizopoulos, Jaakko, et al., 2021b; Naturvårdsverket, 2021). Indicator values for NbS performance can inform decision-makers about the achievement of objectives, or lack thereof, and therefore they become an important tool for the future integration of NbS as climate change mitigation measures (Kumar, Debele, Sahani, Rawat, Marti-Cardona, Alfieri, Basu, Basu, Bowyer, Charizopoulos, Jaakko, et al., 2021b; WMO, 2015).

This chapter aims to present and discuss approaches to assessing the flood regulation services of constructed wetlands at catchment scale. Based on a literature review on the hydrological function of wetlands and the concept of effectiveness, it introduces a modeling approach using the Soil and Water Assessment Tool (SWAT+) and applies it to a case study of the Råån catchment in southwest Sweden.

2 Definition and hydrological functions of wetlands

To understand the role of wetlands as NbS, one must understand the natural systems they are inspired by or intend to mimic. There are many different classification systems of wetlands, and they often incorporate local terms or definitions that are specific to a particular area (Gerbeaux et al., 2018). In Sweden, the most widely accepted wetland definition originates from the Swedish national wetland inventory (VMI), which performed an extensive survey of wetlands in Sweden between 1981 and 2005:

Wetlands are areas where the water table for the main part of the year is close below, at or above the ground level, including vegetation covered lakes. A site is called a wetland when at least 50% of the vegetation is hydrophilic, i.e. water loving. An exception is periodically flooded shores along lakes, seas and rivers, which are classified as wetlands despite a lack of vegetation

Gunnarsson and Löfroth (2009, p. 8).

Unlike their natural counterparts, most constructed wetlands are built for one particular purpose, e.g., the retention of nutrients or to increase water-storage capacity and biodiversity in a catchment. They can be built in different ways, for example, by creating a permanent dam on the main river channel, building a dam for temporary flooding, diverting parts of the river flow to a side dam, or pumping water to an artificial dam (SMHI, 2022).

When the emphasis is placed on the water-storage capacity of constructed wetlands, they become part of a type of green infrastructure called natural water retention measures (NWRM), which are often based on the concept of "keeping the rain where it falls." It is widely accepted that, at catchment scale, implementing such measures in upstream areas can help reduce the risk of downstream flooding (Collentine & Futter, 2018). However, according to Collentine and Futter (2018), implementing NWRM can also involve trade-offs, especially in agricultural areas. When NWRM are installed in upstream rural areas to reduce the risk of flooding in downstream urban areas, it raises questions on costs and benefits between urban and rural interests as well as management and compensation schemes (Collentine & Futter, 2018). The installation of NbS in general requires coordination between many different stakeholders, such as authorities, landowners, and the general public. In Sweden, the integration of NbS has largely taken on a bottom-up approach where landowners and municipalities work together to implement the measures (Interreg North Sea Region, 2020).

Adopting a wetlandscape perspective, which considers the hydrologically coupled system of multiple wetlands and their hydrological catchment (Thorslund et al., 2017), enables investigations on how wetlands are connected to large-scale water fluxes and landscape features such as groundwater and rivers (Åhlén et al., 2020). The cumulative impacts of multiple wetlands in a wetlandscape will depend on their interaction with these water fluxes, and the subsequent fill and spill processes may have an effect on the downstream river hydrograph (Fig. 1).

Typically, it is the proportion of water reservoirs and their volume that has the greatest importance for dampening and retaining water in a catchment area (Naturvårdsverket, 2017). In fact, results from Quin and Destouni (2018) indicate that the flow-dampening effect of floodplain wetlands and lakes at catchment scale is primarily due to their water-storage capacity and less due to their partitioning of precipitation between surface runoff and evapotranspiration. Neither does the interaction of wetlands with underlying groundwater reservoirs seem to be of such a magnitude that it can buffer high water flows (Naturvårdsverket, 2017). However, it is not only the total volume of wetlands

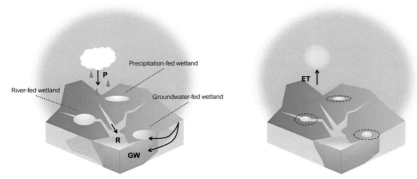

FIG. 1 Conceptual representation of a wetlandscape under wet (left) and dry (right) conditions. The hydrological connection between wetlands and large-scale water fluxes such as precipitation (P), evapotranspiration (ET), runoff (R), and groundwater (GW) flow may alter the downstream river hydrograph. *(Adapted from Thorslund, J., Jarsjo, J., Jaramillo, F., Jawitz, J. W., Manzoni, S., Basu, N. B., Chalov, S. R., Cohen, M. J., Creed, I. F., Goldenberg, R., Hylin, A., Kalantari, Z., Koussis, A. D., Lyon, S. W., Mazi, K., Mard, J., Persson, K., Pietron, J., Prieto, C., Quin, A., Van Meter, K., & Destouni, G., 2017. Wetlands as large-scale nature-based solutions: Status and challenges for research, engineering and management. Ecological Engineering 108, 489–497. http://doi.org/10.1016/j.ecoleng.2017.07.012.)*

that controls their water-storage potential. To provide enough room and responsiveness to rapidly store water and reduce floods, the normal water level in the wetland must be sufficiently low. Antecedent conditions will control the soil moisture and ponding of water in wetlands at the start of a flood event (Acreman & Holden, 2013). For flood mitigation purposes, constructed wetlands must be designed so that there is volume left to fill when high flows occur (Länsstyrelsen Västra Götalands län, 2018).

Still, there is a lack of consensus in the scientific literature on the flow-regulation services of wetlands. Some studies state that wetlands have a large and positive effect on reducing downstream flow, while others conclude that wetlands' ability to regulate flow is negligible or has negative effects on flood mitigation. In an extensive review, Bullock and Acreman (2003) compiled a database of 439 published statements on the water quantity functions of wetlands from 169 studies worldwide. Of these statements, 32 concluded that floods increased or advanced or recession reduced, 62 concluded that floods reduced or delayed or recession increased, and 17 concluded that there was no effect on flood response. In more recent studies, Evenson et al. (2015) found that natural wetlands decrease peak flows and that the reduction was greater following high rainfall events. Yang et al. (2010) concluded that wetland restoration reduces peak discharge, whereas Martinez-Martinez et al. (2014) found that it has a negligible effect on daily peak flow rates and their frequency. The latter study also showed that increasing wetland area, rather than depth, had a greater impact on daily streamflow average. In the Glacial Ridge, where the largest wetland restoration project was implemented in the United States, restoration of wetlands and prairies noticeably reduced downstream flooding. Here, peak floods decreased and flow recessions lengthened, and surface runoff decreased by 33% (Cowdery et al., 2019).

3　Measuring the effectiveness of NbS

3.1　Performance indicators

If constructed wetlands, or any NbS, are to be of practical use in decision-making, one must be able to (with sufficient accuracy) estimate their provided level of service. A wetland can be said to provide a flow-regulation service if, compared to a reference situation (e.g., with and without wetlands), there is a difference in flow regime characteristics (Kadykalo & Findlay, 2016). Kadykalo and Findlay (2016) present a metaanalysis of 28 studies that have investigated the flow-regulation services of wetlands worldwide. In their study, they associate each service with hydrological measurement endpoints or performance indicators (PIs). Examples of indicators of flood reduction services are daily discharge, peak flow, flood frequency, time-to-peak, and peak lag. Thus, PIs are quantitative effectiveness measures, and they relate to factors such as the capacity and function of the system studied (Alegre & Coelho, 2012; Dumitru & Wendling, 2021). NbS effectiveness has, in turn, been defined as "the degree to which objectives are achieved and the extent to which targeted problems are solved. In contrast to efficiency, effectiveness is determined without reference to costs" (Raymond et al., 2017, p. vi).

The NbS handbook developed by representatives of 17 individual EU-funded NbS projects and collaborating institutions (Dumitru & Wendling, 2021) presents a set of indicators to assess the impacts of NbS. Here, the indicators are divided into the following three different types: structural, process-based, and outcome-oriented. The latter refers to NbS performance by measuring baseline conditions (pre-NbS) and evaluating changes in these conditions after NbS implementation. Recommended outcome-based indicators for water management are, for example, flood peak reduction (%), flood peak delay (h), height of flood peak (m^3/s), time to flood peak (h), flood excess volume (m^3), rainfall storage capacity of NbS (mm/%), and surface area of restored and/or created wetlands (ha). Moreover, NbS performance can be evaluated either pre- or postimplementation (Dumitru & Wendling, 2021). However, Kadykalo and Findlay (2016) withhold that there is a limited ability to predict the level of flow-regulation services of wetlands and that ascribing general flow-regulation functions based on current available information is likely unjustified unless detailed in situ data is available.

In the widely quoted manual, *Performance Indicators for Water Supply Services*, Alegre et al. (2016) describe best practices in formulating performance indicators. According to the manual, an indicator should consist of at least two things: (1) a value expressed in specific units and (2) a confidence grade, which indicates the data quality of the indicator (Alegre et al., 2016; Alegre & Coelho, 2012). Ultimately, the goal of any performance indicator is to provide information that can be used for the purpose of making decisions. Thus, performance indicators should not only deliver value but also complementary information such as data quality and context. Context provides information on the inherent characteristics of the system or external events that temporarily affect system performance (e.g., exceptional floods) (Alegre et al., 2016). Furthermore, performance indicators are typically expressed as ratios between variables, and they can be commensurate (e.g., %) or noncommensurate (e.g., €/m³) (Alegre et al., 2016; Alegre & Coelho, 2012). To comply with current best practices, the denominator of noncommensurate ratios should give reference to a system dimension that allows the comparison over time and between systems of different sizes and contexts (Alegre et al., 2016; Santos et al., 2019).

3.2 Modeling wetland interactions at catchment scale

Because in situ observations are often limited to a single wetland and unable to characterize all parameters and processes involved on a catchment scale, hydrologic models can be an effective alternative to assess large-scale wetland functions (Wu et al., 2020). Arnold et al. (2001) list the following three different modeling approaches to design and evaluate constructed wetlands: (1) single event models, (2) stochastic models involving one parameter, and (3) complete water-budget models. One of the most commonly used water-budget models worldwide is the Soil and Water Assessment Tool (SWAT), developed by the USDA Agricultural Research Service and Texas A&M AgriLife Research (Bieger et al., 2017). SWAT is a continuous, conceptual, and semidistributed model proven capable of predicting the impacts of management practices and land use change on water quantity over long periods of time (Gassman et al., 2007; Kumar, Debele, Sahani, Rawat, Marti-Cardona, Alfieri, Basu, Basu, Bowyer, Charizopoulos, Gallotti, et al., 2021a).

Several researchers have used SWAT to assess the cumulative impacts of natural or restored wetlands on catchment hydrology and flood reduction (Evenson et al., 2015; Lee et al., 2018; Wang et al., 2010; Yang et al., 2010). In the Swedish context, the effects of wetlands on catchment or subcatchment hydrology have been studied using the HBV/PULSE model (Johansson, 1993), the HYPE model (Stensen et al., 2019), MIKE FLOOD (DHI, 2020), statistical analysis based on observation data (Quin & Destouni, 2018), as well as wetland survey data coupled with GIS-based analyses (Åhlén et al., 2020; Stengård et al., 2020). If hydrologic models and indicator values are to encourage the undertaking of constructed wetlands, the models need to properly represent the connectivity between man-made hydrologic systems and the natural stream network. Recent advances in the SWAT model have been said to improve such connectivity through added user flexibility (Bieger et al., 2017).

4   Case study of Råån catchment

Råån catchment is located in southwest Sweden, and it covers an area of 193 km² (Fig. 2). The Råån river is 25 km long and originates from Svalöv in the southeast of the catchment. Several tributaries, such as Tjutebäcken, Kövlebäcken, and Lussebäcken, merge with the river before it reaches the sea at Öresund, just south of the city of Helsingborg.

The catchment is located in Sweden's southernmost province, Skåne, whose mixture between coastal and inland climate typically gives rise to strong winds, lower precipitation, and smaller temperature variations compared to other parts of the country (Persson et al., 2011). From 1991 to 2020, the mean annual precipitation in Helsingborg was 666 mm and the mean annual temperature was 8.7°C (SMHI, 2023a). During the same period, the average daily streamflow measured in Råån at the Swedish Meteorological and Hydrological Institute (SMHI) gauging station Bröddebacken was 1.53 m³/s (SMHI, 2023b). The catchment is predominantly rural, and 77% of the catchment area consists of agricultural land (Fig. 3). Urban and other open land stands for about 8% each, while forest makes up around 6% of the area (SMHI, 2023b). According to the National Land Cover Database (Naturvårdsverket, 2019), the total area of wetlands is 153 ha, making up 0.78% of the catchment. Moreover, the predominant soil type in the catchment is clay till and clayey till (Fig. 3), which is characteristic of soils in the Skåne region (Larsson, 2000; SGU, 2020).

The community Råå, which is located where the Råån river meets the sea, is a particularly vulnerable area to flooding due to the combined risk of high water flows in the Råån river and rising sea levels. Heavy rainfall events have already occurred on several occasions, including in 2007, causing high flows in Råån

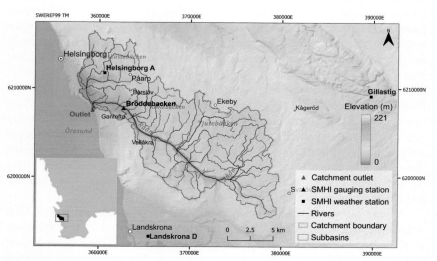

FIG. 2   Location of Råån catchment and its subbasins, rivers, elevation (GSD-Elevation data, grid 50+ nh © Lantmäteriet), as well as SMHI weather and flow gauging stations whose data were considered in this study.

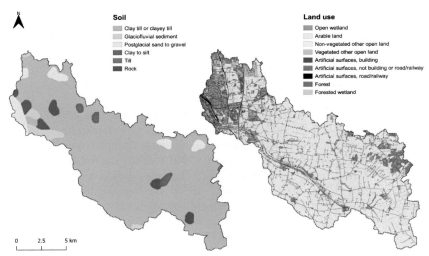

FIG. 3   Soil types (based on Jordarter 1:1 million © SGU) and land use map (National Land Cover Database by Naturvårdsverket) of Råån catchment.

and its tributaries and subsequently cutting off large roads in the area. Even during average maximum flows that recur annually, the land surrounding Råån is at risk of flooding (Helsingborgs stad, 2011).

From 1850 to 1990, the area of natural wetlands within Helsingborg municipality decreased from roughly 30% to 1% (Rååns Vattenråd, n.d.). The drainage of wetlands, the expansion of agricultural land, and the culverting of watercourses have drastically changed water flow dynamics in the Råån catchment (Helsingborgs stad, 2011). Since the 1990s, Helsingborg municipality has implemented various NbS in the catchment. Within the framework of the Interreg project Building with Nature, carried out between 2016 and 2021, a combination of NbS (wetlands, floodplains, stormwater ponds, and two-stage ditches) have been implemented, partially with the aim of reducing flood risk (Naturvårdsverket, 2021). As part of the project, a flood mapping analysis along one of Råån's tributaries, Lussebäcken, was conducted. The aim of the study was to hydraulically model the flood-reducing effects of wetlands, two-stage ditches, and stormwater ponds in the Lussebäcken subcatchment. The results showed that cumulative effects of dams and wetlands had a positive effect on both peak flows and flooding extent for different scenarios of land use and flood return period (DHI, 2020).

Around 60 wetlands have been constructed in the catchment so far, with surface areas ranging from 0.05 to 17 ha (Helsingborgs stad, 2015). Wetlands identified from the National Land Cover Database, including both natural and constructed wetlands, are shown in Fig. 4. The constructed wetlands listed by Rååns Water Board and SMHI's (2023c) constructed wetland database

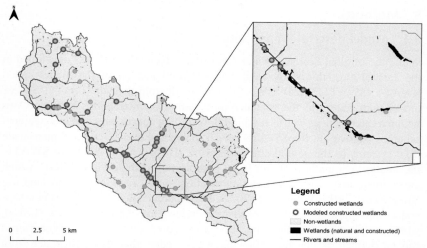

FIG. 4   Wetland distribution (National Land Cover Database from Naturvårdsverket) in Råån catchment and the location of modeled wetlands.

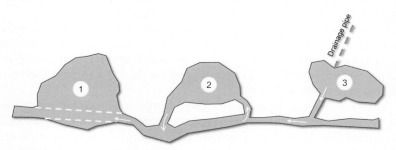

FIG. 5   Three types of constructed wetlands in Råån catchment: (1) wetlands on the channel, (2) side wetlands, and (3) drainage wetlands.

are shown as points in the same figure, also marking the wetlands that were modeled in this study.

There are three different types of constructed wetlands in the catchment, and they can be defined according to their inflow: (1) on-channel wetlands that receive the entire flow from a watercourse; (2) side wetlands that receive only a partial flow from the watercourse, to which both their inlets and outlets are connected; and (3) drainage wetlands that are supplied with water from a culvert or a drainage pipe, e.g., from tile drainage of surrounding farmland (Fig. 5). The water flows in and out of the wetlands either through open channels or via pipes. Typically, the constructed wetlands are shallow (maximum water depth is about 2 m) and naturally formed with flat side slopes (Helsingborgs stad, 2015).

5   Methods

5.1   Model description and setup

While the original Soil and Water Assessment Tool (SWAT) was developed in the early 1990s, this study used a restructured version of the model called SWAT+ (SWAT+, 2024). SWAT+ is more flexible than the original SWAT model with regard to the spatial representation of interactions and processes within a catchment as well as defining flow routing constituents and management schemes. Above all, the SWAT+ model makes it easier to connect man-made flow systems to the natural stream network (Bieger et al., 2017). In SWAT+, the smallest computational unit is the hydrological response unit (HRU), which are lumped areas with a unique combination of soil type, slope, and land use (Bieger et al., 2017; White et al., 2022). During delineation, the catchment is separated into subbasins and subsequently divided into landscape units (LSUs), which are clusters of HRUs, and water areas (Bieger et al., 2017). Water areas can be defined as channels (flowing water bodies that transport water from one point to another), reservoirs (impoundments located on the channel network), and wetlands (water ponding on an HRU) (SWAT+, 2024). A more complete description of the SWAT+ model components and flow paths can be found in Bieger et al. (2017) and in SWAT+ manuals (SWAT+, 2024).

This study used the SWAT+ model version 2020.60.5.5, which was released on March 3, 2023. To set up the catchment, the QGIS interface, QSWAT+, was used. To modify SWAT+ inputs and run the model, the user interface, SWAT+ Editor, was employed. A workflow diagram for the SWAT+ model setup and simulation is shown in Fig. 6.

The SWAT+ model input included data on topography, weather, soil, and land use. A 50 m Digital Elevation Model (DEM) obtained from Lantmäteriet (2022) was used to delineate the boundary of the Råån catchment and its stream network. Two outlets were added to the model: one at the catchment outlet and one at the SMHI gauging station, which is located about 4550 m upstream of the catchment outlet. As a result, the Råån catchment was divided into 15 subcatchments. The daily weather data consisted of precipitation, relative humidity, solar radiation, temperature (min and max), and wind speed from SMHI. To create HRUs, a soil map from the SGU (2020) and a land use map from the National Land Cover dataset provided by the Swedish EPA (Naturvårdsverket, 2019) were used. Both measured and standard values from the literature were used to define the corresponding soil and land use properties (for more details on their preparation, see Dile et al. (2022)). Once the data had been added to the model, the HRUs were created. To ensure as high spatial granularity as possible, no thresholds were set for the HRU definition, which resulted in the creation of 2476 HRUs.

5.2   Modeling wetlands in SWAT+

There are two main ways that wetlands can be represented in SWAT+: (1) as reservoirs on channels, and (2) as isolated reservoirs located in the landscape, off the channel network. In SWAT+, the first type is referred to as a "reservoir,"

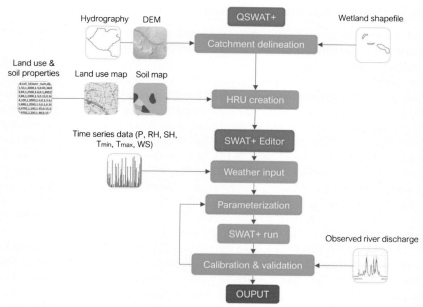

FIG. 6 Workflow diagram of SWAT+ model setup and simulation. DEM is Digital Elevation Model, HRU is hydrological response unit, P is precipitation (mm), T is temperature (°C), RH is relative humidity (in fraction), WS is wind speed (m/s), and SR is solar radiation (MJ/m^2/day).

and the second type is referred to as a "wetland." This study used both types to model wetlands in the Råån catchment. All areas that are specified in the land use map as wetlands are automatically registered as such in SWAT+ Editor. Since the constructed wetlands are connected to the river network, they were modeled as "reservoirs," and most attention (parameterization, calibration, etc.) was paid to the reservoir type.

In SWAT+, reservoirs are added manually to the catchment, before delineation, as polygon shapefiles. In this study, the boundaries of the polygon shapes were traced according to the water surface of the constructed wetlands visible on satellite imagery. During delineation, the channel that is intersected by a reservoir polygon is split into an upstream and downstream portion. The user can then reconnect the channels by routing portions of the flow from the upstream channel to the wetland and/or downstream channel. Drainage wetlands, on the other hand, receive inflow from tile drainage in the surrounding farmland via pipes and culverts. In SWAT+, tile drains can be added to any HRU. The exact location and dimensions of the tile drains were unknown in this study, which is why standard values for Swedish soil conditions from the Swedish Board of Agriculture were used (Larsson et al., 2013). In their wetland catalog, Helsingborgs stad (2015) provides the size of the subcatchment area that drains into each constructed wetland. In this study, tile drains were added to agricultural HRUs with roughly the same area as the ones specified in the catalog, and the drainage hydrograph was routed to the wetlands.

5.3   Wetland data and parameterization

An accurate representation of the constructed wetlands requires knowledge of their inflow, outflow, area, and volume. Due to limited data, 35 out of the 60 wetlands were modeled as reservoirs in this study. The others were assumed to contribute with some storage capacity as "wetland modules" from the land use map, although their connection to the river network was not represented. In their wetland catalog from 2015, the municipality of Helsingborg provides the year of construction and original maximum depth of several constructed wetlands in the Råån catchment as well as information on their types of inflow and outflow (Helsingborgs stad, 2015). In Lumsén (2021), estimated values of inflow to the constructed wetlands from the Råån Water Board are given. Lumsén (2021) also provides estimates of sediment accumulation in 28 constructed wetlands in the Råån catchment (0.2–2.3 cm/year), which were used in this study to parametrize the new depths of the wetlands. A list of the 35 modeled wetlands and available data are shown in Table 1.

The reservoirs in SWAT+ have the following two conceptual storage zones: a principal (normal) and an emergency (maximum) volume. The volume of the constructed wetlands was calculated using an equation that relates area and depth to wetland volume. Hayashi and Van der Kamp (2000) used morphometric data to develop simple equations to represent the volume-area-depth relations of shallow wetlands. Brooks and Hayashi (2002) improved these equations to also include maximum depth. The Eq. (1) for maximum volume of a wetland (V_{max}) includes maximum area (A_{max}), maximum depth (d_{max}), and a basin shape coefficient called p (Brooks & Hayashi, 2002).

$$V_{max} = \frac{A_{max} \times d_{max}}{1 + (2/p)} \tag{1}$$

In this study, d_{max} was assigned the original maximum depth provided by Helsingborgs stad (2015) minus the sedimentation rate (cm/year) estimated by Lumsén (2021) multiplied by the age of the wetland. The p-coefficient was arbitrarily set to 1.5.

5.4   Wetland water balance and outflow simulation

The water balance for a reservoir in SWAT is shown in Eq. (2) below:

$$V = V_{stored} + V_{flowin} - V_{flowout} + V_{pcp} - V_{evap} - V_{seep}, \tag{2}$$

where V is the volume of water in the reservoir at the end of the day, V_{stored} is the volume of water stored at the beginning of the day, V_{flowin} is the volume of water entering the reservoir at the beginning of the day, $V_{flowout}$ is the volume of water flowing out of the water body during the day, V_{pcp} is the volume of precipitation falling on the water body, V_{evap} is the volume of water removed from the reservoir due to evapotranspiration, and V_{seep} is the volume lost by seepage from the bottom of the reservoir (Neitsch et al., 2011).

TABLE 1 List of the 35 modeled constructed wetlands and available data included in this study.

Name*	Type*	Year of construction*	Max depth (m)*	Wetland area (m²)	Catchment area (ha)*	Inflow (m³/year)**
R1: Tjutebro	D	1991	1.5	3307	25	71,200
R2: Raus	S	1992	0.8	6490	16,300	
R3: Gantofta	S	1992	0.5	3422	14,800	10,539,100
R4: Fastmårup	S	1992	1.2	7193	11,500	
R5: Pålstorp-Gantofta	S	1992	0.5	14,248	26.5	
R5: Pålstorp-Gantofta	S	1992	0.5	3175	26.5	
R6: Ormastorp N	D	1993	2	10,480	52	
R7: Ormastorp S	D	1993	2	1438	229	
R8A: Bälteberga A	S	1993	1.5	2059	8600	
R8c: Bälteberga C1	S	1993	1	661	8600	1,224,800
R8c: Bälteberga C2	S	1993	1	2046	8600	1,224,800
R8c: Bälteberga C3	S	1993	1	2151	8600	
R8d: Bälteberga D	S	1993	1	1247	8600	1,224,800
R8e: Bälteberga E	S	1994	1.5	973	8600	
R10: Ramlösa	C	1994	2.5	5743	900	2,563,600
R11: Ekeberga 2	S	1994	1	1295	200	
R11: Ekeberga 3	S	1994	1	1580	200	
R14: Stenbrogården	D	1994	1.4	5309	147	
R17: Ormastorp	D	1995	2	1144	25	71,200
R18: Strömsnäs	S	1995	2.2	6870	8500	6,052,900

Continued

TABLE 1 List of the 35 modeled constructed wetlands and available data included in this study—cont'd

Name*	Type*	Year of construction*	Max depth (m)*	Wetland area (m²)	Catchment area (ha)*	Inflow (m³/year)**
R19: Rönnebäck	C	1995	3	1078	420	299,100
R21: Vasatorp	C	1995	2	6443	500	1,424,200
R23: Rönnedals gård	D	1995	2	782	7	
R24: Södergårds kvarndamm	D	1996	2	1408	230	
R25: Kvistofta	D	1996	1.5	1512	50	142,400
R29: Kingelstad V	D	1997	2.1	2611	40	113,900
R30: Ljungberga	S	1997	1.9	871	200	569,700
R32a: Södra Vallåkra A	D	1997	0.9	2246	40	
R32c: Södra Vallåkra C	D	1997	2	1341	40	113,900
R34: Kingelstad Ö	D	1999	3.5	1900	120	340,500
R35: Norra Vallåkra	S	2001	1.3	2096	916	1,166,800
R36: Fågelsjön	–	2002	1.5	23,686	75	
R37: Kingelstad N	S	2002	2.1	8279	2100	1,545,300
R38: Långeberga försöksanläggning	S	2003	1	1742	850	
R40: Bunketofta	D	2007	2	7199	350	

Type D=drainage wetland; S = side wetland; and C = wetland-on-channel. Columns marked with * are data from Helsingborgs stad (2015); ** is from the Råån Water Board in Lumsén (2021), and the wetland area originates from the reservoir polygons created in QGIS.

The SWAT+ reservoir and wetland routine allows the user to determine $V_{flowout}$ with different release actions. Using decision tables, the user can, for example, input a measured release rate, a weir equation, or drawdown days (Arnold et al., 2018). A reservoir or wetland releases water at a specified rate whenever the reservoir or ponded volume exceeds the principal spillway volume, and the rate is determined by the total number of days during which the excess volume is released (Neitsch et al., 2011). When the volume exceeds the emergency spillway volume, the excess water is released at a higher rate.

5.5 Simulation and sensitivity analysis

The SCS Curve Number method (SCS, 1972) was used to simulate the precipitation-runoff processes, and a warm-up period of 2 years was used to establish initial hydrological conditions. To evaluate the flood regulation services of the constructed wetlands, the assessment was based on a paired simulation scenario: with and without wetlands (i.e., 100% removal of the constructed wetlands). The former constituted the "baseline model" and was subject to calibration and validation. The latter was created by making a copy of the baseline model, so that all other parameters, HRUs, etc., remained the same. The water was then re-routed past the side and on-channel wetlands, and instead of routing the tile flow to the drainage wetlands, it was routed to the nearest channel. These changes were then evaluated against the baseline model to assess changes in the downstream hydrograph.

In SWAT, reservoirs and wetlands may have a large influence on the river hydrograph. This fact, combined with uncertainties in wetland dimensions, made it necessary to assess the impact of reservoir and wetland parameters on simulated flow. A summary of the changes made during the sensitivity analysis is shown in Table 2.

TABLE 2 Changes in wetland HRUs and reservoirs during the sensitivity analysis.

Objects and variables	Change
Replacement of wetland HRUs with agricultural land	100% replacement
Removal of side and on-channel wetlands	100% removal
Removal of side, on-channel, and drainage wetlands	100% removal
Principal and emergency area and volume	Percent changes
Principal and emergency release days	Release days decreased

5.6 Model calibration and validation

The model was calibrated and validated against observed streamflow at a daily time step. It is important to choose a relatively stable calibration period, which should succeed in the years when the constructed wetlands have been implemented. Because the land use data in this study were produced between 2017 and 2019, the calibration was chosen to span across 3 hydrological years starting in 2017 (October 1, 2017 to September 30, 2020), and the validation period consisted of the 2 subsequent years of record (October 1, 2020 to December 31, 2022). Average daily streamflow records from the SMHI station Bröddebacken were used to calibrate model parameters. Because this station is not located at the catchment outlet, the daily streamflow simulated by the SMHI model S-HYPE at the outlet was also compared to the simulated flow. Furthermore, data on the estimated inflow to the wetlands provided by the Råån Water Board were used to calibrate the fraction of flow routed to the modeled wetlands.

Model performance was evaluated by visually comparing the daily river hydrograph of observed and simulated flow over the calibration period and using the Nash-Sutcliffe efficiency (NSE) (Eq. 3). NSE has proven to reflect the overall fit of the hydrograph well, although it has a bias toward errors in high flows (Moriasi et al., 2007).

$$
\text{NSE} = 1 - \left[\frac{\sum_{i=1}^{n} \left(Q_i^{\text{obs}} - Q_i^{\text{sim}} \right)^2}{\sum_{i=1}^{n} \left(Q_i^{\text{obs}} - Q^{\text{mean}} \right)^2} \right]
\tag{3}
$$

In this case, Q_i^{obs} is the ith observation of streamflow at the gauging station, Q_i^{sim} is the ith simulated value of streamflow, Q^{mean} is the mean of the observed streamflow data, and n is the total number of observations. NSE values range between $-\infty$ and 1.0 (optimal), and according to Moriasi et al. (2007), model performance can be judged as satisfactory if the monthly NSE value is above 0.5. For finer resolution time steps (e.g., daily), a lower threshold is warranted (Evenson et al., 2015).

5.7 Choosing performance indicators

The choice of performance indicators will depend on the explicit objective of the catchment area and the NbS implementation program. This study did not connect specific objectives to wetland properties and indicators. Rather, it suggests a selection of indicators that are possible (but not limited to) to evaluate using the outputs from the SWAT+ model and that encompass varied aspects that have been shown to be important in NbS implementation.

The level of flood regulation services of the constructed wetlands was partially evaluated by comparing the characteristics of the daily hydrograph. Characteristics of the flow regime, such as timing, peak flow rate, and recession curve shape, were visually assessed across the calibration period. Some of the proposed indicators in previous studies do not focus on the outcome but rather on describing the capacity of the system. Therefore, the average storage capacity of the constructed wetlands was included as an indicator. The storage volume of the wetlands needed to be put in relation to a wetland dimension (e.g., depth, volume, and surface area), as stressed by Santos et al. (2019). Since the volumes were calculated using a theoretical volume-area-depth equation, and Martinez-Martinez et al. (2014) showed that wetland surface area has a greater impact on streamflow than depth, the storage capacity was evaluated in cubic meters per square meter of wetland area.

As highlighted by Alegre et al. (2016), it is important that performance indicators are ascribed with a confidence grade and a context, i.e., inherent system characteristics or external factors (e.g., rain events). In the case of the hydrograph characteristics, the NSE value indicated the data quality of the indicators. Because the impact on daily streamflow averages over long periods of time can balance out possible differences in high and low flow events, the average daily streamflow for different ranges of daily precipitation was analyzed. Moreover, the maximum flow in a given year does not necessarily have to overtop the confines of the river channel. A more consequential type of performance indicator would therefore be to directly assess overbank flooding from the channels. Lastly, according to the Dumitru and Wendling (2021), any NbS impact model evaluation needs a social acceptance factor to measure aspects such as transparency, communication, awareness, and citizen engagement. In this study, "social acceptance" was only broadly evaluated by assuming the conclusions from the Interreg project Building with Nature (Interreg North Sea Region, 2020). A summary of the performance indicators that were chosen in this study and the corresponding methods used to evaluate them are shown in Table 3.

TABLE 3 Selection of performance indicators and the method used to evaluate them.

Indicator	Unit	Method
Peak flow rate	–	Hydrograph
Time-to-peak	–	Hydrograph
Recession curve shape	–	Hydrograph
Event-based	m^3/s	Daily average streamflow for different ranges of daily precipitation
Daily average streamflow	m^3/s	SWAT+ output file
Average wetland storage capacity	m^3/ha or m^2	SWAT+ output file
Overbank flooding from channels	mm	SWAT+ output file
Social acceptance	–	Interreg project Building with Nature

6   Results and discussion

6.1   Model performance

For the 3-year calibration period, the simulated average daily streamflow at the gauging station ($1.27 \, m^3/s$) was slightly underestimated compared to the observed streamflow during the same period ($1.35 \, m^3/s$). The overall model evaluation statistics (Table 4) suggest that the model performance is within the range of previous SWAT and SWAT+ studies. Considering that the model exceeded the stricter monthly performance criteria of $NSE = 0.5$ during both the calibration and validation period, the model can be judged satisfactory.

A comparison between daily simulated and observed streamflow at the SMHI gauging station Bröddebacken during the calibration and validation period is shown in Fig. 7. The model was able to reproduce low flow during the dry summer season well but was less capable of reproducing peak flows during the flood season. As expected in most hydrologic models, the accuracy of the simulated hydrograph during the validation period was lower than during calibration. This discrepancy suggests that with the current catchment configuration, the model was not able to adequately represent all hydrological processes in the catchment. Generally, the validation period experienced heavier rainfall and higher peak flow rates, which again points to the model's inability to reproduce high-flow events.

The corresponding comparison using the SMHI S-HYPE model at the catchment outlet is shown in Fig. 8. The simulated hydrograph shows signs of flood peak attenuation compared to the flow measured at the gauging station. While it may be expected that the peak flows gradually decrease when a flood propagates in the downstream direction, it is uncertain how the estimates from the S-HYPE model would compare to observed data at the outlet. This, in turn, makes it difficult to draw any definite conclusions on this study's model performance using the S-HYPE model. Nevertheless, the model was able to estimate the timing (rising time, duration, and recession) of the peak flows well. During the validation period, the S-HYPE model also showed higher model performance than the gauged data (0.80 vs 0.58). Again, this is likely due to the fact that peak flows

TABLE 4 Summary of model performance.

Station	Period	Daily NSE
SMHI Bröddebacken	Calibration	0.73
	Validation	0.58
SMHI S-HYPE at outlet	Calibration	0.84
	Validation	0.80

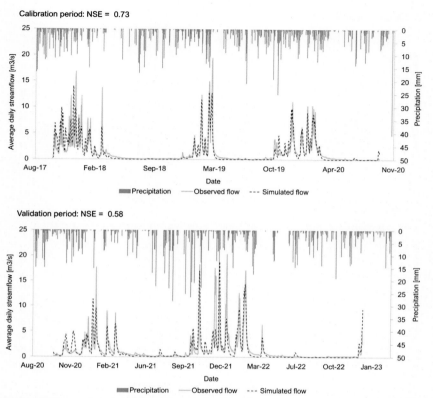

FIG. 7 Comparison between daily simulated and observed streamflow at the SMHI gauging station Bröddebacken during calibration (top) and validation (bottom) periods.

were generally lower in the S-HYPE model, and NSE is known to have a high sensitivity to errors in high flows due to the square in the residuals.

6.2 Sensitivity of wetland parameters

The model showed a general insensitivity to changes in wetland dimensions (area and volume) as well as the complete removal of wetland and reservoir modules. These results differ from Yang et al. (2010), whose SWAT model was very sensitive to changes in the normal and maximum wetland areas and volumes. The complete removal of the wetland HRUs and reservoirs had no visible effects on the river hydrograph or model performance. The model did, however, show large sensitivity toward changes in the release mechanisms of the constructed wetlands, more specifically to the number of emergency release days (Fig. 9).

The shape of the recession curve can give information on the retention characteristics of a catchment, as it is often governed by the release of water from

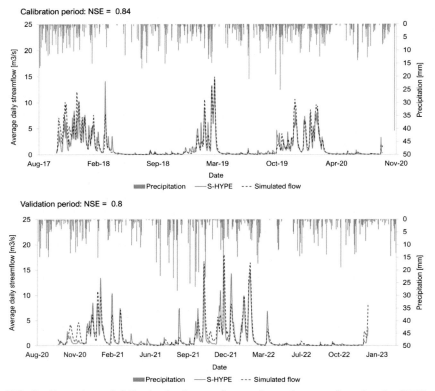

FIG. 8 Comparison of daily simulated streamflow at the catchment outlet using the SMHI S-HYPE model during calibration (top) and validation (bottom).

natural storage (Jain, 2011). When the constructed wetlands were allowed to release excess water at a slower rate, the recession rates in the simulated hydrograph decreased and the peak flows were reduced. The calibrated model did, however, have the highest NSE value when the excess water was immediately spilled downstream at the same time increment as the model (daily), causing no attenuation in flood peaks—just as if the wetlands were not there. Moreover, the model showed no sensitivity to changes in the principal release days, which suggests that the maximum volume of the wetlands was frequently exceeded in the model. Indeed, the water balance indicated that some wetlands frequently reached their maximum storage capacity such that high-flow events, routing large quantities of water to the reservoirs, immediately induced spillage to the downstream channel.

Due to the lack of in situ wetland water balance monitoring data, proper validation of simulated wetland hydrology and parameters (area and volume) was not undertaken in this study. Should long-term daily inflow and outflow data become available, future studies could better calibrate and validate catchment-wetland

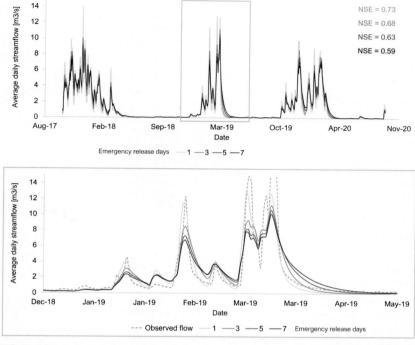

FIG. 9   Model sensitivity toward changes in emergency release days across the entire calibration period (top) and across a portion of the hydrograph (bottom).

interactions at shorter timescales. Moreover, there was an uncertainty in reservoir-specific release rules and dimensions of the outlet structures of the wetlands, which compromised the accuracy of the reservoir release rates specified in the decision tables.

6.3   Assessment of wetland effectiveness

This study considered a system of performance indicators that gave reference to context, system dimensions, and confidence grades that may be used to explain the performance of the modeled wetlands and overall model output (Fig. 10).

Because the removal of the constructed wetlands had a negligible effect on the downstream hydrograph, the associated performance indicators (peak flow rate, time-to-peak, and recession curve shape) were ascribed as having no impact. Compared to findings from previous studies, Yang et al. (2010) concluded that wetland conservation and restoration reduce peak discharge, whereas Martinez-Martinez et al. (2014) instead found that wetland restoration had a negligible effect on daily peak flow rates. Percentwise, the impact on daily average streamflow was a mere 0.02% reduction in this study. In terms of the

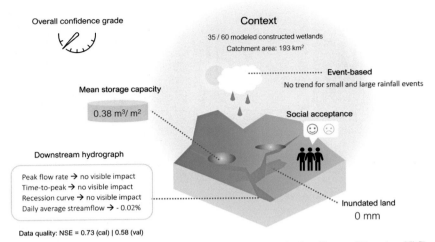

FIG. 10 System of performance indicators evaluated by comparing baseline conditions (pre-NbS) to changes in the indicators after the implementation of 35 constructed wetlands.

TABLE 5 Daily average streamflow at the gauging station across the calibration period for different ranges of daily precipitation, with and without constructed wetlands.

Daily precipitation ranges (mm/day)	<5	5–10	10–20	>20
Flow without constructed wetlands (m^3/s)	1.2086	2.3787	3.2697	0.5347
Flow with constructed wetlands (m^3/s)	1.2083	2.3789	3.2682	0.5353
Percent difference (%)	−0.0258	0.0063	−0.0453	0.1245

event-based indicator, there was no trend shown for changes in streamflow with increasing size of rainfall event (Table 5). Compared to findings in previous studies, Evenson et al. (2015) concluded that the reduction in peak flows of natural wetlands was greater following high rainfall events compared to smaller events, and Wu et al. (2020) found flood attenuation to be larger for high than low flows.

The constructed wetlands had an average storage capacity of $0.38\,m^3$ per m^2 of wetland area. However, this storage capacity seemed to have little influence on the overall catchment hydrology as the constructed wetlands allowed high flows to immediately pass through to the downstream channel due to the daily emergency release rate. Furthermore, the model exhibited no overbank flooding from channels during the entire calibration period. This is likely not the case in reality, as accounts from Helsingborg municipality suggested even annual flow peaks tend to flood banks around the Råån river. This result could be due to the general underestimation of flood peaks in the model or due to inaccurate representations of channel dimensions.

Regarding the social acceptance indicator, the support from farmers in the Råån catchment was crucial in implementing the NbS measures, and a "bottom-up" approach encouraged collaborations between municipalities and landowners, who offered their land without gaining economic compensation (Interreg North Sea Region, 2020). This speaks for a general acceptance toward recently constructed wetlands in the Råån catchment, but quantifying a proper social acceptance factor would require more detailed investigations of the public's perception. As was stressed by Collentine and Futter (2018), placing water retention measures on rural land to mitigate flood risk in urban areas could lead to trade-offs. Although the constructed wetlands typically do not claim large areas of land, they can still lead to income losses if placed on productive farmland. In fact, the Building with Nature Project concluded that upscaling NbS in the Råån catchment will require both new legislation and economic instruments to compensate landowners.

In terms of data quality, the performance of the hydrological model was satisfactory, but the lack of in situ wetland flow data and dimensions motivated a low overall confidence grade for the performance indicators. Regarding context, the primary functions of the 35 wetlands were registered as nutrient retention and increase of biodiversity. Although it is likely that constructed wetlands that are not explicitly adapted to reduce flow will still have such a function, they will likely be less effective in doing so. This stresses the need to coordinate individual wetland functions with catchment-scale objectives. Indeed, for successful implementation of constructed wetlands as NbS, the focus must not only lie on isolated projects and local site conditions, but planning must be done at wetlandscape scale (Ferreira et al., 2022). Moreover, the borders of catchments typically do not follow administrative boundaries, and it is necessary that management decisions are coordinated accordingly. In the neighboring catchment, Höje å, a collaborative project between 3 municipalities (Lund, Lomma, and Staffanstorp) resulted in the restoration of more than 80 wetlands, with the aim of reducing eutrophication, promoting biodiversity, and reducing the risk of flooding in downstream areas. While the issue of flooding still remains, the follow-up has shown good results regarding the eutrophication problem, and the municipalities intend to continue to work holistically, with each other and landowners, to implement solutions further up in the catchment area (Naturvårdsverket, 2017).

While it is evident that planning must be conducted at catchment scale, it is important to realize the constraints of catchment models in assessing the effectiveness of wetlands. Typically, the larger the model domain, the more input data are required, and the higher becomes the complexity and uncertainty of the model. In addition, the cumulative impact of multiple wetlands at catchment scale can differ significantly from the functions and effects observed at local wetland scales, which raises the question of what scales to assess their effectiveness. In this study, the flood regulation services of constructed wetlands showed to have limited impact, while the DHI (2020) study of the smaller subcatchment Lussebäcken concluded that constructed wetlands and dams had a positive effect on both peak flows and flooding extent. It is evident, then, that predictions at subcatchment scale could give a very different idea on the effectiveness of the wetlands. With smaller model domains and knowledge of the local area, perhaps they also have the potential to be more accurate. In the end, the choice of model and appropriate scale will depend on the region and NbS objectives. As Thorslund et al. (2017) also conclude, investigations of wetlandscapes will likely require a combination of different methods, such as statistical approaches, analytical modeling, ground-based measurements, remote sensing, and GIS-based analysis.

7 Concluding remarks and future research needs

The flow-regulating capability of the constructed wetlands was directly related to their outflow mechanisms. To what extent this capability was utilized during each flood event likely depended on the antecedent water level and the magnitude of the flow entering the wetlands. In this study, the constructed wetlands seemed to frequently surpass their maximum storage capacity, causing any additional water to get spilled immediately downstream. Since the model exhibited the best performance when the wetlands released excess water during the same daily time increment as the model, the modeled wetlands seemed to provide limited flood regulation services. However, as other authors have emphasized, ascribing general flow-regulation functions and services to wetlands is likely unjustified unless detailed in situ monitoring data are available.

Ultimately, the choice of appropriate performance indicators will depend on the objective of the NbS implementation program, and it is important that individual wetland functions and catchment-scale objectives align. Although flood analyses are mostly concerned with peak flows, a comprehensive set of performance indicators may also encompass other characteristics of the flow regime such as time-to-peak, recession rates, and duration. In order to facilitate comparisons over time and between systems of different sizes and contexts, it is important that the performance indicators give reference to system dimensions. To increase transparency, the performance indicators should also be ascribed with contextual information and a confidence grade. However, the usefulness of the indicators in decision-making will fully depend on the quality of the data used to derive them. If indicator values are to be useful tools in the future integration of NbS as climate change mitigation measures, long-term monitoring data of the solutions are required.

Future modeling approaches to evaluating wetland effectiveness need to focus on accurately portraying wetland water balance components, which likely requires collaborations with local authorities and landowners to install in situ monitoring stations. Because this study had a high sensitivity toward the release mechanisms of the constructed wetlands, future SWAT+ studies should experiment with different release rules and create customized decision tables that better represent the outlet structures of the studied wetlands. While this study focused on retrospective impact evaluation of wetlands, future research work could investigate the potential of using SWAT+ in the planning phase of constructed wetlands (design, location, etc.) using a catchment-scale perspective. The added flexibility and user-friendly GIS-based interface could make it suitable as a sort of decision support tool, but as other authors have highlighted, it can be difficult to predict wetland performance under future management schemes.

Acknowledgments

This study was developed as part of the project PUDDLE JUMP: Promoting Upstream-Downstream Directed Linkages in the Environment: "Joined-Up" Management Perspectives, funded by FORMAS (reference 2022-02138). Carla Ferreira thanks the national funding by FCT—Foundation for Science and Technology, P.I., through the institutional scientific employment program contract (CEECINST/00077/2021).

References

Acreman, M. C., & Holden, J. (2013). How wetlands affect floods. *Wetlands*, *33*(5), 773–786. https://doi.org/10.1007/s13157-013-0473-2.

Adikari, Y., & Yoshitani, J. (2009). *Global trends in water-related disasters: An insight for policymakers*. UNESCO, International Centre for Water Hazard and Risk Management (ICHARM). https://unesdoc.unesco.org/ark:/48223/pf0000181793 (Accessed 23.02.02).

Åhlén, I., Hambäck, P., Thorslund, J., Frampton, A., Destouni, G., & Jarsjö, J. (2020). Wetlandscape size thresholds for ecosystem service delivery: Evidence from the Norrström drainage basin, Sweden. *Science of the Total Environment*, *704*, 135452. https://doi.org/10.1016/j.scitotenv.2019.135452.

Alegre, H., Baptista, J. M., Cabrera, E., Cubillo, F., Duarte, P., Hirner, W., Merkel, W., & Parena, R. (2016). *Performance indicators for water supply services—Manual of best practices* (3rd ed.). London, UK: IWA Publishing.

Alegre, H., & Coelho, S. T. (2012). Infrastructure asset management of urban water systems. In A. Ostfeld (Ed.), *Water supply system analysis* (pp. 49–73). IntechOpen. https://doi.org/10.5772/52377.

Arnell, N. W., & Gosling, S. N. (2016). The impacts of climate change on river flood risk at the global scale. *Climatic Change*, *134*(3), 387–401. https://doi.org/10.1007/s10584-014-1084-5.

Arnold, J. G., Allen, P. M., & Morgan, D. S. (2001). Hydrologic model for design and constructed wetlands. *Wetlands*, *21*(2), 167–178. https://doi.org/10.1672/0277-5212(2001)021[0167:HMFDAC]2.0.CO;2.

Arnold, J. G., Bieger, K., White, M. J., Srinivasan, R., Dunbar, J. A., & Allen, P. M. (2018). Use of decision tables to simulate management in SWAT+. *Water*, *10*(6), 713. https://doi.org/10.3390/w10060713.

Bieger, K., Arnold, J. G., Rathjens, H., White, M. J., Bosch, D. D., Allen, P. M., Volk, M., & Srinivasan, R. (2017). Introduction to SWAT+, a completely restructured version of the soil and water assessment tool. *JAWRA Journal of the American Water Resources Association, 53*(1), 115–130. https://doi.org/10.1111/1752-1688.12482.

Brooks, R. T., & Hayashi, M. (2002). Depth-area-volume and hydroperiod relationships of ephemeral (vernal) forest pools in southern New England. *Wetlands, 22*, 247–255. https://doi.org/10.1672/0277-5212(2002)022[0247:DAVAHR]2.0.CO;2.

Bullock, A., & Acreman, M. (2003). The role of wetlands in the hydrological cycle. *Hydrology and Earth System Sciences, 7*(3), 358–389. https://doi.org/10.5194/hess-7-358-2003.

Collentine, D., & Futter, M. N. (2018). Realising the potential of natural water retention measures in catchment flood management: Trade-offs and matching interests. *Journal of Flood Risk Management, 11*(1), 76–84. https://doi.org/10.1111/jfr3.12269.

Cowdery, T. K., Christenson, C. A., & Ziegeweid, J. R. (2019). The hydrologic benefits of wetland and prairie restoration in western Minnesota-lessons learned at the glacial ridge national wildlife refuge, 2002-15. In *Scientific investigations report* US Geological Survey. 2019–5041 https://pubs.usgs.gov/sir/2019/5041/sir20195041.pdf (Accessed 23.04.14).

Davidson, N. (2014). How much wetland has the world lost? Long-term and recent trends in global wetland area. *Marine and Freshwater Research, 65*, 934–941. https://doi.org/10.1071/MF14173.

DHI. (2020). *Flood analysis Lussebäcken, Helsingborg—Hydraulic impact of urbanization and effect of implemented water management measures.* https://docplayer.se/197057134-Flood-analysis-lussebacken-helsingborg.html (Accessed 23.02.02).

Dile, Y., Srinivasan, R., & George, C. (2022). *QGIS interface for SWAT+: QSWAT+ Version 2.3.* https://swatplus.gitbook.io/docs/user/qswat+ (Accessed 23.04.14).

Dumitru, A., & Wendling, L. (2021). *Evaluating the impact of nature-based solutions: A handbook for practitioners.* European Commission, Directorate-General for Research and Innovation, Publications Office of the European Union. https://data.europa.eu/doi/10.2777/244577 (Accessed 23.04.14).

European Commission, Directorate-General for Research and Innovation, & Vojinovic, Z. (2020). *Nature-based solutions for flood mitigation and coastal resilience: Analysis of EU-funded projects.* Publications Office of the European Union. https://data.europa.eu/doi/10.2777/374113 (Accessed 23.02.02).

Evenson, G. R., Golden, H. E., Lane, C. R., & D'Amico, E. (2015). Geographically isolated wetlands and watershed hydrology: A modified model analysis. *Journal of Hydrology, 529*, 240–256. https://doi.org/10.1016/j.jhydrol.2015.07.039.

Feng, B., Zhang, Y., & Bourke, R. (2021). Urbanization impacts on flood risks based on urban growth data and coupled flood models. *Natural Hazards, 106*, 613–627. https://doi.org/10.1007/s11069-020-04480-0.

Ferreira, C. S. S., Kalantari, Z., Hartmann, T., & Pereira, P. (2022). Introduction: Nature-based solutions for flood mitigation. In C. S. S. Ferreira, Z. Kalantari, T. Hartmann, & P. Pereira (Eds.), *Nature-based solutions for flood mitigation—Environmental and socio-economic aspects* Springer Nature Switzerland AG. https://doi.org/10.1007/698_2021_776. 1-10 pp.

Ferreira, C. S. S., Kašanin-Grubin, M., Solomun, M. K., Sushkova, S., Minkina, T., Zhao, W., & Kalantari, Z. (2023). Wetlands as nature-based solutions for water management in different environments. *Current Opinion in Environmental Science & Health, 33*, 100476. https://doi.org/10.1016/j.coesh.2023.100476.

Gassman, P. W., Reyes, M. R., Green, C. H., & Arnold, J. G. (2007). The soil and water assessment tool: Historical development, applications, and future research directions. *Transactions of the ASABE, 50*(4), 1211–1250. https://doi.org/10.13031/2013.23637.

Gerbeaux, P., Finlayson, C. M., & van Dam, A. A. (2018). Wetland classification: Overview. In C. Finlayson, M. Everard, K. Irvine, R. J. McInnes, B. A. Middleton, A. van Dam, & N. Davidson (Eds.), *The wetland book I: Structure and function, management and methods* (pp. 1461–1468). Dordrecht: Springer. https://doi.org/10.1007/978-90-481-9659-3_329.

Gunnarsson, U., & Löfroth, M. (2009). *Våtmarksinventeringen—Resultat från 25 års inventeringar, Nationell slutrapport för våtmarksinventeringen (VMI) i Sverige. Vol. 5925.* Naturvårdsverket, ISBN:978-91-620-5925-5. https://www.naturvardsverket.se/om-oss/publikationer/5900/vatmarksinventeringen- -resultat-fran-25-ars-inventeringar/ (Accessed 23.04.14) (in Swedish).

Hayashi, M., & Van der Kamp, G. (2000). Simple equations to represent the volume–area–depth relations of shallow wetlands in small topographic depressions. *Journal of Hydrology, 237* (1–2), 74–85. https://doi.org/10.1016/S0022-1694(00)00300-0.

Helsingborgs stad. (2011). *PM Klimatanpassning.* https://media.helsingborg.se/uploads/networks/4/sites/141/2014/11/PM_Klimatanpassning_KF.pdf (Accessed 23.04.16) (in Swedish).

Helsingborgs stad. (2015). *Anlagda våtmarker—Våtmarker, tvåstegsdiken och dagvattendammar.* https://media.helsingborg.se/uploads/networks/1/2015/08/vatmarker_katalog_2015_low_sbf.pdf (Accessed 23.04.14) (in Swedish).

Interreg North Sea Region. (2020). *Nature based solutions demand better coordination in catchments.* https://northsearegion.eu/media/12482/skaane-policy-brief-catchment-20200220-1.pdf (Accessed 23.05.01).

Jain, M. K. (2011). Recession of discharge. In V. P. Singh, P. Singh, & U. K. Haritashya (Eds.), *Encyclopedia of earth sciences series. Encyclopedia of snow, ice and glaciers.* Dordrecht: Springer. https://doi.org/10.1007/978-90-481-2642-2_437.

Johansson, B. (1993). *Modelling the effects of wetland drainage on high flows. Vol. RH 8.* SMHI. https://www.smhi.se/publikationer/modelling-the-effects-of-wetland-drainage-on-high-flows-1.7173 (Accessed 23.02.02).

Kadykalo, A. N., & Findlay, C. S. (2016). The flow regulation services of wetlands. *Ecosystem Services, 20,* 91–103. https://doi.org/10.1016/j.ecoser.2016.06.005.

Kreplin, H. N., Ferreira, C. S. S., Destouni, G., Keesstra, S. D., Salvati, L., & Kalantari, Z. (2021). Arctic wetland system dynamics under warming. *WIREs Water, 8,* e1526. https://doi.org/10.1002/wat2.1526.

Kumar, P., Debele, S. E., Sahani, J., Rawat, N., Marti-Cardona, B., Alfieri, S. M., Basu, B., Basu, A. S., Bowyer, P., Charizopoulos, N., Gallotti, G., Jaakko, J., Leo, L. S., Loupis, M., Menenti, M., Mickovski, S. B., Mun, S.-J., Gonzalez-Ollauri, A., Pfeiffer, J., … Zieher, T. (2021). Nature-based solutions efficiency evaluation against natural hazards: Modelling methods, advantages and limitations. *Science of the Total Environment, 784*(147058). https://doi.org/10.1016/j.scitotenv.2021.147058.

Kumar, P., Debele, S. E., Sahani, J., Rawat, N., Marti-Cardona, B., Alfieri, S. M., Basu, B., Basu, A. S., Bowyer, P., Charizopoulos, N., Jaakko, J., Loupis, M., Menenti, M., Mickovski, S. B., Pfeiffer, J., Pilla, F., Proell, J., Pulvirenti, B., Rutzinger, M., … Zieher, T. (2021). An overview of monitoring methods for assessing the performance of nature-based solutions against natural hazards. *Earth-Science Reviews, 217*(103603). https://doi.org/10.1016/j.earscirev.2021.103603.

Länsstyrelsen Västra Götalands län. (2018). *Naturanpassade åtgärder mot översvämning—Ett verktyg för klimatanpassning.* 2018:13 https://www.lansstyrelsen.se/download/18.2887c5dd16488fe880d5ad3b/1537797458169/Naturanpassade%20%C3%A5tg%C3%A4rder%20mot%20%C3%B6versv%C3%A4mning.pdf (Accessed 23.04.14) (in Swedish).

Lantmäteriet. (2022). *Terrain model download, grid 50+ NH. Document version 1.6.* https://www.lantmateriet.se/globalassets/geodata/geodataprodukter/hojddata/e_grid50plus_nh.pdf (Accessed 23.04.21).

Larsson, R. (2000). *Lermorän—En litteraturstudie.* Swedish Geotechnical Institute. Varia 480 https://www.diva-portal.org/smash/get/diva2:1300384/FULLTEXT01.pdf (Accessed 23.04.14) (in Swedish).

Larsson, T., de Maré, L., Lindmark, P., Rangsjö, C.-J., & Johansson, T. (2013). *Jordbrukets markavvattningsanläggningar i ett nytt klimat.* Jordbruksverket. 2013:14 https://www2. jordbruksverket.se/webdav/files/SJV/trycksaker/Pdf_rapporter/ra13_14.pdf (Accessed 23.04.14).

Lee, S., Yeo, I.-Y., Lang, M. W., Sadeghi, A. M., McCarty, G. W., Moglen, G. E., & Evenson, G. R. (2018). Assessing the cumulative impacts of geographically isolated wetlands on watershed hydrology using the SWAT model coupled with improved wetland modules. *Journal of Environmental Management, 223,* 37–48. https://doi.org/10.1016/j.jenvman.2018.06.006.

Lumsén, L. (2021). *Analys av sedimentackumulation i våtmarker i Rååns avrinningsområde och undersökning av våtmarkernas renoveringsbehov.* https://raan.se/wp-content/uploads/2021/ 03/Louise_Lumsen_examensarbete.pdf (Accessed 23.04.14) (in Swedish).

Martinez-Martinez, E., Nejadhashemi, A. P., Woznicki, S. A., & Love, B. J. (2014). Modeling the hydrological significance of wetland restoration scenarios. *Journal of Environmental Management, 133,* 121–134. https://doi.org/10.1016/j.jenvman.2013.11.046.

Moriasi, D. N., Arnold, J. G., Van Liew, M. W., Bingner, R. L., Harmel, R. D., & Veith, T. L. (2007). Model evaluation guidelines for systematic quantification of accuracy in watershed simulations. *Transactions of the ASABE, 50*(3), 885–900. https://swat.tamu.edu/media/1312/ moriasimodeleval.pdf (Accessed 23.05.14).

Naturvårdsverket. (2017). *Kunskapsunderlag om våtmarkers ekologiska och vattenhushållande funktion.* NV-05712-17 http://vatmarksguiden.se/wp-content/uploads/2018/02/NV_2017.pdf (Accessed 23.04.14) (in Swedish).

Naturvårdsverket. (2019). *Nationella marktäckedata 2018 basskikt. Utgåva 1.0.* https://wiki. openstreetmap.org/w/images/8/8e/NMD_Produktbeskrivning_NMD2018Basskikt_v1_0.pdf (Accessed 23.04.26) (in Swedish).

Naturvårdsverket. (2021). *Naturbaserade lösningar—Ett verktyg för klimatanpassning och andra samhällsutmaningar. Vol. 7016.* https://www.naturvardsverket.se/globalassets/media/ publikationer-pdf/7000/978-91-620-7016-2.pdf (Accessed 23.02.02) (in Swedish).

Neitsch, S. L., Arnold, J. G., Kiniry, J. R., & Williams, J. R. (2011). *Soil and water assessment tool theoretical documentation version 2009.* Texas Water Resources Institute Technical Report No. 406 https://swat.tamu.edu/media/99192/swat2009-theory.pdf (Accessed 23.04.14).

Persson, G., Sjökvist, E., Åström, S., Eklund, D., Andréasson, J., Johnell, A., Asp, M., Olsson, J., & Nerheim, S. (2011). *Klimatanalys för Skåne län.* SMHI. No 2011-52 https://www. lansstyrelsen.se/download/18.2e0f9f621636c84402730f3d/1528811635925/LSTM-SMHI_2012_ Klimatanalys%20f%C3%B6r%20Sk%C3%A5ne%20l%C3%A4n.pdf (Accessed 23.04.14) (in Swedish).

Quin, A., & Destouni, G. (2018). Large-scale comparison of flow-variability dampening by lakes and wetlands in the landscape. *Land Degradation & Development, 29*(10), 3617–3627. https:// doi.org/10.1002/ldr.3101.

Rååns Vattenråd, n.d. Våtmarker. https://raan.se/?page_id=825 (Accessed 23.04.16) (in Swedish).

Raymond, C. M., Berry, P., Breil, M., Nita, M. R., Kabisch, N., de Bel, M., Enzi, V., Frantzeskaki, N., Geneletti, D., Cardinaletti, M., Lovinger, L., Basnou, C., Monteiro, A., Robrecht, H., Sgrigna, G., Munari, L., & Calfapietra, C. (2017). An impact evaluation framework to support planning and evaluation of nature-based solutions projects. In *EKLIPSE expert working group on nature-based solutions to promote climate resilience in urban areas.* https://doi.org/ 10.13140/RG.2.2.18682.08643.

Santos, L. F., Galvão, A. F., & Cardoso, M. A. (2019). Performance indicators for urban storm water systems: A review. *Water Policy, 21*(1), 221–244. https://doi.org/10.2166/wp.2018.042.

SCS. (1972). *National engineering handbook. Section 4: Hydrology.* U.S. Department of Agriculture.

SGU. (2020). *Kartvisaren Jordarter 1:1 miljon.* https://www.sgu.se/produkter-och-tjanster/kartor/ kartvisaren/jordkartvisare/jordarter-11-miljon/ (Accessed 23.04.14) (in Swedish).

SMHI. (2022). *Anlagda våtmarker.* https://www.smhi.se/kunskapsbanken/hydrologi/vatmarker/ anlagda-vatmarker-1.178384 (Accessed 23.04.21) (in Swedish).

SMHI. (2023a). *Dataserier med normalvärden för perioden 1991–2020.* https://www.smhi.se/data/ meteorologi/dataserier-med-normalvarden-for-perioden-1991-2020-1.167775   (Accessed 23.04.14) (in Swedish).

SMHI. (2023b). *Modelldata per område. Delavrinningsområdets SUBID 254.* http://vattenwebb. smhi.se/modelarea/ (Accessed 23.04.14) (in Swedish).

SMHI. (2023c). *Anlagda våtmarker.* https://vattenwebb.smhi.se/wetlands/ (Accessed 23.04.16) (in Swedish).

Stengård, E., Räsänen, A., Ferreira, C. S. S., & Kalantari, Z. (2020). Inventory and connectivity assessment of wetlands in northern landscapes with a depression-based DEM method. *Water, 12,* 3355. https://doi.org/10.3390/w12123355.

Stensen, K., Matti, B., Rasmusson, K., & Hjerdt, N. (2019). *Modellstudie för att undersöka åtgärder som påverkar lågflöden—Delrapport 2 i regeringsuppdrag om åtgärder för att motverka vattenbrist i ytvattentäkter.* SMHI. No. Hydrologi 121 https://www.smhi.se/publikationer/ modellstudie-for-att-undersoka-atgarder-som-paverkar-lagfloden-delrapport-2-i-regeringsuppd rag-om-atgarder-for-att-motverka-vattenbrist-i-ytvattentakter-1.152550 (Accessed 23.02.02) (in Swedish).

SWAT+. (2024). *SWAT+ IO documentation.* https://swatplus.gitbook.io/docs (Accessed 23.04.18).

Thorslund, J., Jarsjo, J., Jaramillo, F., Jawitz, J. W., Manzoni, S., Basu, N. B., Chalov, S. R., Cohen, M. J., Creed, I. F., Goldenberg, R., Hylin, A., Kalantari, Z., Koussis, A. D., Lyon, S. W., Mazi, K., Mard, J., Persson, K., Pietron, J., Prieto, C., ... Destouni, G. (2017). Wetlands as large-scale nature-based solutions: Status and challenges for research, engineering and management. *Ecological Engineering, 108,* 489–497. https://doi.org/10.1016/j.ecoleng.2017.07.012.

Vojinovic, Z., Alves, A., Gomez, J. P., Weesakul, S., Keerakamolchai, W., Meesuk, V., & Sanchez, A. (2021). Effectiveness of small- and large-scale nature-based solutions for flood mitigation: The case of Ayutthaya, Thailand. *Science of the Total Environment, 789*(147725). https://doi. org/10.1016/j.scitotenv.2021.147725.

Wang, X., Shang, S., Qu, Z., Liu, T., Melesse, A. M., & Yang, W. (2010). Simulated wetland conservation-restoration effects on water quantity and quality at watershed scale. *Journal of Environmental Management, 91*(7), 1511–1525. https://doi.org/10.1016/j. jenvman.2010.02.023.

White, M. J., Arnold, J. G., Bieger, K., Allen, P. M., Gao, J., Čerkasova, N., Gambone, M., Park, S., Bosch, D. D., Yen, H., & Osorio, J. M. (2022). Development of a field scale SWAT+ modeling framework for the contiguous US. *JAWRA Journal of the American Water Resources Association, 56*(6), 1545–1560. https://doi.org/10.1111/1752-1688.13056.

WMO. (2015). *Effectiveness of flood management measures. Integrated Flood Management Tools Series No. 21 version 1.0.* https://www.floodmanagement.info/publications/tools/Tool_21_ Effectiveness_of_Flood_Management_Measures.pdf (Accessed 23.04.14).

Wu, Y., Zhang, G., & Rousseau, A. N. (2020). Quantitative assessment on basin-scale hydrological services of wetlands. *Science China Earth Sciences, 63*(2), 279–291. https://doi.org/10.1007/ s11430-018-9372-9.

Yang, W., Wang, X., Liu, Y., Gabor, S., Boychuk, L., & Badiou, P. (2010). Simulated environmental effects of wetland restoration scenarios in a typical Canadian prairie watershed. *Wetlands Ecology and Management, 18*(3), 269–279. https://doi.org/10.1007/s11273-009-9168-0.

Chapter 1.4

A new framework for urban flood volume estimation using low-impact development methods and intelligent models

Yashar Dadrasajirlou[a], Hojat Karami[a], Alireza Rezaei[a], Seyedali Mirjalili[b], Zahra Kalantari[c], and Carla Sofia Santos Ferreira[d,e]

[a]Civil Engineering Department, Semnan University, Semnan, Iran, [b]Artificial Intelligence Research and Optimization Centre, Torrens University, Adelaide, SA, Australia, [c]KTH Royal Institute of Technology, Stockholm, Sweden, [d]Polytechnic Institute of Coimbra, Applied Research Institute, Coimbra, Portugal, [e]Research Centre for Natural Resources Environment and Society (CERNAS), Polytechnic Institute of Coimbra, Coimbra, Portugal

1 Introduction

Urban areas are home to more than half of the world's population (UN, 2012). The rapid growth and concentration of population in the cities comprise a relevant challenge for managing the impacts of natural hazards and in particular the risk of flooding (Ferreira et al., 2016; Flood, 1997). Floods are one of the most well-known devastating natural disasters in the world (Freer et al., 2013; Razavi et al., 2020). Increasing urbanization rates in the world have caused impermeable surfaces to replace permeable soils (Zhang et al., 2015). The spread of impermeable areas parallel with climate change has caused a considerable increase in flood volume in urban areas (Olang & Furst, 2011; Zhang et al., 2018). In contrast, the occurrence of heavy rainfall challenges the urban drainage system and thus has major impacts on communities, habitats, and the economy (Li et al., 2020). The direct effect of urban development on the increase of surface runoff has been mentioned in several studies (Dietz & Clausen, 2008; Ferreira et al., 2012; Palla & Gnecco, 2015; Schoonover et al., 2006; Wang et al., 2005). The impacts of urban areas on storm runoff are variable and can affect both the quality and quantity of runoff (Huang et al., 2018; Jacobson, 2011; Shen et al., 2014). Managing the volume of storm runoff with urban drainage systems is a common challenge for planners due to urban flooding (Bisht et al., 2016). Especially in arid and semiarid regions,

Nature-Based Solutions in Supporting Sustainable Development Goals
https://doi.org/10.1016/B978-0-443-21782-1.00006-3

83

managing storm runoff volume by enhancing local water retention has received increasing attention (Bond et al., 2014; Ghazavi et al., 2012; Huang et al., 2018; Teng et al., 2017).

In recent years, new methods for flood control in urban areas have been explored by researchers, such as those based on nature-based solutions (Ferreira et al., 2020). Nature-based solutions, known as green infrastructure, have been widely recommended for reducing runoff from urban flooding, improving water quality, and having a sustainable ecosystem (Fu et al., 2019). It has been reported that among the various methods of urban flood management, green infrastructure, best management practices, and low-impact development methods (LIDs) have interesting performance in reducing runoff and peak discharge (Eckart et al., 2012; Hoang & Fenner, 2016; Jia et al., 2013; Wolch et al., 2014). LID reduces the volume and speed of stormwater runoff and decreases the impacts of floods (Ahiablame et al., 2012; Kayhanian et al., 2012; Randhir & Raposa, 2014; Teymouri et al., 2020), but their design optimization is still challenging (Babaei et al., 2018; Geng & Sharpley, 2019; Martin-Mikle et al., 2015). The effectiveness of green infrastructure in reducing urban flooding is affected by a combination of several factors, such as local environmental settings, the distribution of green infrastructure, and the hydraulic characteristics of the green infrastructure (Fiori & Volpi, 2020; Park et al., 2013). The complexity of the urban environment requires integrated research and strategies to control urban flooding (Lawson et al., 2014; Pappalardo et al., 2017).

The optimal design of runoff management systems is based on mathematical models (Faye et al., 2016). Various modeling methods have been used to predict flooding (Henonin et al., 2015; Mark et al., 2004; Miguez et al., 2017; Rene et al., 2014). The Storm Water Management Model (SWMM) is one of the most widely used rainfall-runoff modeling methods for modeling flood and peak discharge in urban catchments (Shahed Behrouz et al., 2020). In recent years, intelligent models have been used extensively in predicting hydrological and hydraulic processes. A few examples are the Support Vector Machine (SVM) used to predict flood susceptibility (Choubin et al., 2018; Karami et al., 2022; Lei et al., 2021; Panahi et al., 2022) and evaporation from open water surfaces (Aghelpour et al., 2021), the Least Square Support Vector Machine (LSSVM) to predict runoff (Azad et al., 2021; Liu et al., 2021) and drought (Pham et al., 2021), and the combination of variational mode decomposition and the Least Squares Support Vector Machine optimized by the Sparrow Search Algorithm (VMD-SSA-LSSVM) to predict water level dynamics (Song et al., 2021). Although intelligent algorithms have been used for flood prediction, their application in assessing the impact of LIDs is rather limited. This chapter proposes a new method to find the best percentage cover of LIDs used in urban areas to mitigate floods in urban areas, based on SVM, LSSVM, and LSSVM-GOA algorithms.

2 Materials and methods

2.1 Study area

Golestan town ($\sim$2.86 km^2) is located in the northern part of Semnan City, in Semnan province (Fig. 1). The city of Semnan is located in the north of the desert plain and south of the Alborz Mountain range. Its average altitude is 1132 m above sea level. Golestan town comprises the largest number of residential settlements within the city of Semnan. The maximum and minimum altitudes of the town are 1244 and 1175 m, respectively.

2.2 Methodology

The methodological framework used in this study is summarized in Fig. 2 and explained in the following subsections.

2.2.1 Low-impact development methods

Storm water is usually water that does not penetrate the ground but flows on the ground (including impermeable surfaces) into streams, rivers, lakes, etc. The low-impact development methods used to reduce flood volume and peak discharge investigated in this study include rain barrels (RB), permeable pavement (PP), and infiltration trench (IT). These methods (Fig. 3) have been selected according to the arid and semiarid climate conditions of the region. The specifications of LIDs considered for this study are summarized in Table 1.

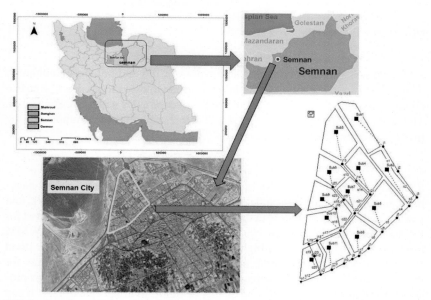

FIG. 1 Location of Golestan town case study in Semnan City and Semnan Province in Iran.

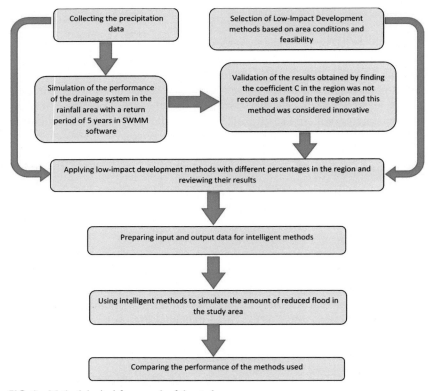

FIG. 2 Methodological framework of the study.

Seven scenarios involving the three types of LIDs and/or their combination were investigated regarding flood mitigation in Golestan town. Each scenario contains 10 plans, i.e., different percentage cover of LIDs within the study area. The scenarios investigated are listed in Table 2, where a few examples of the investigated plans are also provided.

2.2.2 SWMM model

SWMM is a dynamic model for rainfall-runoff simulation, urban flood management and planning, analysis and design of surface water collection network, and sewage and drainage systems in the urban basins. This model was presented for the first time in 1971 by the US Environmental Protection Agency. SWMM model is easy to use and has high power in quantitative and qualitative flood simulation. SWMM is a single event or continuous model that accounts for the evaporation, snowmelt, seepage, deep infiltration, and subsurface flow processes to simulate the quantity and quality of runoff in urban catchments. It can record the flow in three modes: permanent flow, kinematic wave, and dynamic wave (Tavakol-Davani et al., 2016).

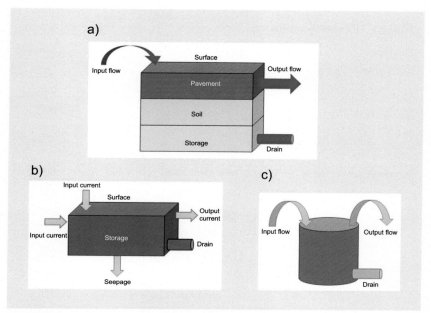

FIG. 3 Schematic representation of functioning of the LID methods investigated in this study: (A) permeable pavement, (B) infiltration trench, and (C) rain barrel.

We used the SWMM model to investigate the runoff volume and peak discharge in Golestan case study. The model uses hydrological data measured at the outlet of the basin (for more details, please read Dadrasajirlou et al., 2023). Due to the lack of historical flood records in the area, the model was calibrated and validated only with the peak flow obtained from the software and compared with the peak flow calculated from the logical relation (Eq. 1) in each of the subcatchments. Considering that each subcatchment is covered by different land uses, the area occupied by individual land uses in each subcatchment was calculated using AutoCAD software. Then, by using relative weight averaging (Eq. 1) to the area of each subcatchment, the runoff coefficient of each subcatchment was calculated.

$$\overline{C} = \frac{\sum_{i=1}^{n} C_i A_i}{\sum_{i=1}^{n} A_i} \tag{1}$$

C_i is the runoff coefficient related to the area A_i of each subcatchment.

After calculating the $\overline{C}$ value for each subcatchment using the logical formula (Eq. 2), the peak flow rate in each subbasin was calculated.

TABLE 1 LID specifications.

LID type	LID features	Amount
Rain barrel	Barrel height (mm)	1000
	Flow coefficient	0.8
	Flow exponent	0.5
	Offset (mm)	6
	Daily drain (h)	6
Permeable pavement	Surface roughness	0.12
	Thickness (mm)	120
	Void ratio	0.6
	Flow coefficient	0.8
	Flow exponent	0.5
	Offset (mm)	15
	Soil conductivity (mm/h)	120.4
	Soil thickness (mm)	150
	Storage thickness (mm)	250
Infiltration trench	Surface slope (%)	1
	Storage thickness (mm)	1000
	Void ratio	0.75
	Flow coefficient	0.8
	Offset (mm)	6

$$Q = 0.278 * C * i * A \qquad (2)$$

where Q stands for maximum peak discharge (m^3/s), C is the runoff coefficient, i is the rainfall intensity (mm/h), and A is the area of the catchment (km^2).

The results of the SWMM model are used to train and validate the machine learning models.

2.2.3 Intelligent algorithms

In this study, we have selected three intelligent algorithms to optimize the modeling process to investigate the impact of LIDs on flood mitigation in the study site. We selected the SVM, LSSVM, and LSSVM-GOA algorithms based on their highest performance and accuracy for flood assessment, considering the

TABLE 2 LID scenarios investigated in this study.

Scenario	Type	Symbol	Example	Explanation of example
1	Individual	RB-percentage	RB-30	The LID used is a rain barrel that covers 30% of the study area
2	Individual	PP-percentage	PP-20	The LID used is a permeable pavement, covering 20% of the study area
3	Individual	IT-percentage	IT-70	The LID used is an infiltration trench, covering 70% of the study area
4	Paired composition	IT-PP-percentage	IT-PP-35	The LIDs used are a combination of permeable pavement and infiltration trench, and the coverage of each of them is 35%
5	Paired composition	IT-RB-percentage	IT-RB-15	The LIDs used are infiltration trench and rain barrel, and the percentage cover of each of them is 15%
6	Paired composition	PP-RB-percentage	PP-RB-45	The LIDs used are permeable pavement and rain barrel, and the percentage used of each of them is 35%
7	Triple composition	IT-PP-RB-percentage	IT-PP-RB-20	It involves the use of the three types of LIDs, and the percentage used by each of them is 20%

available literature. In the following subsections, we introduce each of the models. We also provide a short description of the Grasshopper Optimization Algorithm (GOA) given its integration with the LSSVM algorithm.

The input data used for the intelligent models include the percentage of LIDs and K coefficient. To use the K coefficient, it was necessary to consider the rate

TABLE 3 Statistical details of input data used for modeling.

	Input			
	Rain barrel coverage (%)	Porous pavement coverage (%)	Infiltration trench coverage (%)	$K_{Q\text{-}peak}$
Max	100	100	100	0.782363977
Min	0	0	0	0.786116323
Average	18.286	18.286	18.286	0.818198874

of peak flow reduction associated with each mode compared to the non-LID mode. The study area includes 12 subareas (Fig. 2). The range of data used is shown in Table 3. The maximum flood volume and maximum runoff for the outfall point were considered as model outputs.

2.2.3.1 Support Vector Machine

Vapnik introduced a support vector machine in 1995 (Vapnik, 1995). The biggest positive feature of this algorithm is that it does not fall into the trap of local optimization using global optimization methods in its structure. The SVM algorithm also maps the input vector to a higher dimensional space using a nonlinear function. By using linear regression, SVM calculates the amount of output. It is assumed that (x, y) are observational period data, where x is the input vector and y is the observational output. Eq. (3) formulates the linear regression as follows:

$$y' = f(x) = \omega^T \phi(x) + b \tag{3}$$

where $f(x)$ represents the linear relationship, and $\phi(x)$ represents the nonlinear mapping function. ω and b represent the weight and bias. To reduce the difference between the model outputs and the actual outputs, it is necessary to minimize Eq. (4):

$$\min : \psi = \frac{1}{2}\|\omega\|^2 + \gamma \sum_{i=1}^{n}\left(\xi_i + \xi_i^*\right)$$

$$\text{subject to} : \begin{cases} \omega\phi(x_i) + b - y \leq \varepsilon + \xi_i \\ y - \omega\phi(x_i) + b \leq \varepsilon + \xi_i^* \\ \xi_i, \xi_i^*, i = 1, 2, 3..., n \end{cases} \tag{4}$$

In the first term of this relationship, $\frac{1}{2}\|\omega\|^2$ is the norm of the weights. The second term, γ, is a real positive number showing the penalty coefficient, ξ_i and ξ_i^* are also the penalty coefficients of the upper and lower error. The accuracy of the model is indicated by the ε parameter. Therefore, model simplicity and experimental error are indicated by the first and second expressions, respectively. In this research, the radial kernel function has been used (Eq. 5).

$$K(x, x_i) = \exp\left(\frac{-\|x - x_i\|^2}{2\sigma^2}\right) \tag{5}$$

where σ indicates the width of the kernel function, and K represents the nonlinear kernel function.

2.2.3.2   Least Square Support Vector Machine

Suykens first introduced the LSSVM, which is characterized by its relatively short computation time achieved through the optimization method used (Suykens, 2001). In other words, the nonlinear relationship between inputs and outputs becomes a linear relationship by considering the mapping from lower dimension space to higher dimension space (Suykens, 2001). This is one of the most practical methods of solving nonlinear problems (Anandhi et al., 2008). The linear relation expressed in LSSVM is formulated in Eq. (6):

$$y' = W^T\Phi(x) + b \tag{6}$$

where W is the weight of the inputs, b is the bias, Φ is the mapping on linear function, x is the input, and y is the output.

By minimizing Eq. (7), the difference between modeled data and real data is reduced to its lowest value (Anandhi et al., 2008):

$$\text{Min}: \Psi(W, e) = \frac{1}{2}W^*W^T + \frac{1}{2}C\sum_{i=1}^{N} e_i \tag{7}$$

$$\text{Subject to}: e_i = y_i - y_i'$$

In Eq. (7), the coefficient C is used to apply the penalty coefficient and has a positive and fixed value. If the value of coefficient C is high, it causes the complexity of the equation, and if its value is low, it reduces the complexity of the equation. The first part of Eq. (7) is used to show the norm of weights. In other words, the value of this part is directly related to the complexity of the problem, so if its value decreases, it means that the problem is less complex.

2.2.3.3   Grasshopper Optimization Algorithm

GOA is an intelligent optimization method introduced by Saremi et al. (2017) to solve complex engineering problems. This algorithm has advantages compared to other optimization algorithms, such as the participation of all search agents in updating the position of each search agent, special attention to avoid falling into the trap of local optimization and convergence, and balancing global and local search capabilities (Saremi et al., 2017). In this algorithm, the position of the grasshoppers in the group is defined as follows:

$$X_i = S_i + G_i + A_i \tag{8}$$

In this regard, S_i is the effect of social interaction, G_i is the effect of gravity, and A_i is the effect of wind on the movement of the grasshopper. Eq. (9) is used to simulate the social interaction of grasshopper:

$$S_i = \sum_{j=1}^{n} s(d_{ij}) d'_{ij} \tag{9}$$

where d_{ij} represents the Euclidean distance between the ith and the jth grasshoppers, s is the stress due to the community force, and d_{ij}' is the unit vector, which indicates the direction of the movement of the ith grasshopper toward the jth grasshopper. The function is defined as follows:

$$S(r) = fe^{-r/l} - e^{-r} \tag{10}$$

where f is the absorption intensity and l is the absorption length scale. In this equation, two forces of absorption and repulsion are considered. When the distance of the grasshoppers from each other is between 0 and 2.079, the force of them is repulsive, and when this distance is between 2.079 and 4, this force is gravitational. Also, if the distance is 2.079, no force is created (Saremi et al., 2017). Eq. (11) shows the formula for updating the position of the grasshoppers:

$$X_i^d = c \left[\sum_{j=1}^{n} c((ub_d - lb_d)/2) s \left(\left| x_j^d - x_i^d \right| \right) \frac{x_j - x_i}{d_{ij}} + T_d \right] \tag{11}$$

In this relation, ub and lb represent the upper limit and the lower limit of the decision variables, respectively. T indicates the goal or best position achieved so far. Parameter c represents a decreasing coefficient to limit the neutral zone, repulsion zone, and gravity zone. The c parameter causes the group of grasshoppers to converge toward the target. This position, which is followed by a group of grasshoppers, will be updated if a new position is found. Eq. (12) represents the formula for updating the parameter in each iteration:

$$c = c_{max} - l(c_{max} - c_{min})/L \tag{12}$$

In this regard, L represents the maximum number of iterations, l the current number of iterations, $c_{max} = 1$, and $c_{min} = 0.00001$.

2.2.3.4 Least Square Support Vector Machine-Grasshopper Optimization Algorithm

The hybrid LSSVM-GOA algorithm is described as follows (Fig. 4):

(1) The basic parameters of GOA such as population, frequency, and number of repetitions are determined
(2) Preparing training data and test data
(3) Production of the initial population
(4) Training LSSVM and decision variables of GOA using training data
(5) Test LSSVM and determine the target function of GOA

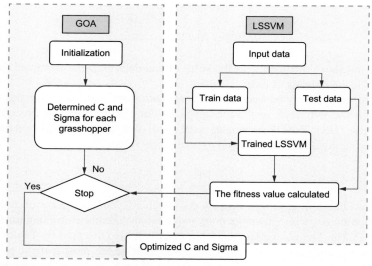

FIG. 4   Flowchart of LSSVM-GOA algorithm.

(6) Check the termination conditions; if the termination conditions are satisfied, return the optimal values of the LSSVM parameters; otherwise, change the position of each Grasshopper and repeat steps 4 and 5.

2.2.4   Statistical indicators

To determine the accuracy of runoff volume reduction (VR) predicted by the intelligent models, evaluation indicators including Coefficient of Determination (R^2), Mean Square Error (MAE), and Root Mean Square Error (RMSE) are used, following Eqs. (13), (14), (15), respectively. The coefficient of determination indicates the degree of correlation between the real data (in this case, the output of the SWMM model) and the modeled data from each intelligent model; values close to one indicate a good correlation. The MAE and RMSE indices show the rates of error, and values close to zero indicate more precise predictions (Karami et al., 2022).

$$R^2 = 1 - \left[\frac{\sum_{i=1}^{n}(E_i - G_i)^2}{\sum_{i=1}^{n}(E_i)^2 - \frac{\sum_{i=1}^{n}(G_i)^2}{N}} \right] \tag{13}$$

$$\text{MAE} = \frac{1}{N}\sum_{i=1}^{n}|E_i - G_i| \tag{14}$$

$$\text{RMSE} = \sqrt{\frac{\sum_{i=1}^{n} (E_i - G_i)^2}{N}} \qquad (15)$$

In the above relations, E_i is the value of real data, G_i is the estimated value, and N is the number of data.

The comparison between models was based on their accuracy but also through the DRGraph method, a graph layout algorithm by dimensionality reduction (DR). The DR parameter is the result of dividing the predicted values by the intelligent model (G_i) into experimental values (E_i):

$$\text{DR} = \frac{G_i}{E_i} \qquad (16)$$

The DRGraph generates visually comparable layouts by means of a sparse distance matrix, with a lower memory requirement and shorter running time (Zu et al., 2021). The comparison between models is also assessed by Taylor diagrams, a mathematical diagram designed to indicate the most realistic model. This graph is plotted to compare the results of each algorithm by comparing the value of their standard deviation, correlation coefficient, and root mean square error.

3 Results and discussion

3.1 Hydrological results based on SWMM model

The results included in this section are the result of the hydrological simulation of SWMM software.

3.1.1 Model performance

Fig. 5 shows the value of the R^2 between the peak flow values calculated with the logical formula (Eq. 1) and the peak flow simulated with the SWMM for the 5-year return period. According to $R^2 = 0.8657$ and RMSE $= 0.01653 \, \text{m}^3/\text{s}$, it can be concluded that the SWMM model has a good ability to simulate runoff generation in the study area.

3.1.2 Hydrological impact of LIDs

Fig. 6 provides a graphical representation of the impact of seven scenarios and associated plans with the highest and lowest percentage cover on the hydrograph. Based on the scenarios focusing on individual LID, RB performs better than IT and PP in flood reduction. The performance of IT and PP scenarios is similar, although PP provides a slightly higher flood reduction. The RB scenario shows an increasing flood reduction potential with increasing LID coverage, ranging from 74% to 79.5% of runoff volume reduction and 79% to 83% of peak

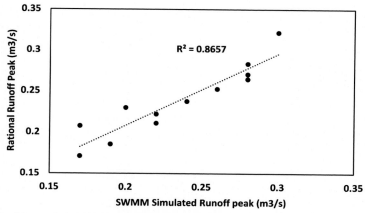

FIG. 5 Linear relationship between the peak discharge obtained from the logical equation and peak runoff estimated using the SWMM model.

discharge reduction for RB-10 and RB-100, respectively. The results of the paired scenarios reveal that the simultaneous use of IT and PP can provide a reduction of up to 90% of the peak discharge and 70% of the flood volume when considering IT-PP-45 and IT-PP-50 plans. However, the combined use of RB and PP provides the best flood mitigation results from all seven scenarios, with the largest reduction of peak discharge (80%–98%) and reduction in flood volume (70%–90%). The RB scenario, the IT-RB scenario, and the IT-PP-RB scenario are next in terms of peak discharge reduction, respectively. The IT scenario revealed the weakest performance in terms of flood mitigation.

3.2   Performance of the intelligent models

3.2.1   Accuracy of the models

Fig. 7 shows the results of volume reduction (VR) prediction in both training and validation stages for SVM, LSSVM, and LSSVM-GOA models. In the training stage, the LSSVM-GOA model revealed a higher accuracy ($R^2 = 0.9874$, MAE $= 0.0124\,\mathrm{m^3/s}$ and RMSE $= 0.0209\,\mathrm{m^3/s}$) than the SVM ($R^2 = 0.948$, MAE $= 0.0234\,\mathrm{m^3/s}$ and RMSE $= 0.0422\,\mathrm{m^3/s}$) and the LSSVM ($R^2 = 0.8977$, MAE $= 0.0418\,\mathrm{m^3/s}$ and RMSE $= 0.0594\,\mathrm{m^3/s}$) models. In the validation process, LSSVM-GOA model ($R^2 = 0.9896$) also outperformed the LSSVM ($R^2 = 0.9266$) and SVM ($R^2 = 0.8990$), providing results with high accuracy. Also, in this process, LSSVM-GOA model showed higher predictive power with values of 0.0101 and 0.0185 for MAE and RMSE indices, respectively, compared to LSSVM model with values of 0.0268 and 0.0361 and SVM model with values of 0.0318 and 0.0434.

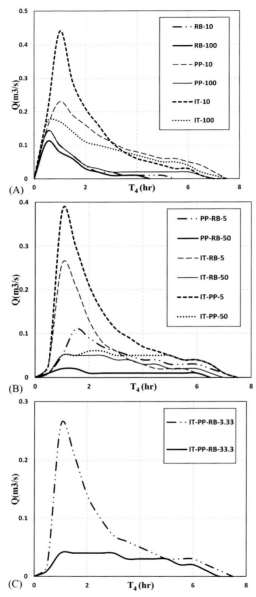

FIG. 6 Impact of different LID scenarios on the 5-year return period hydrographs: (A) individual scenarios, (B) pair scenarios (combination of two LIDs), and (C) triple composition scenario (combination of three LIDs).

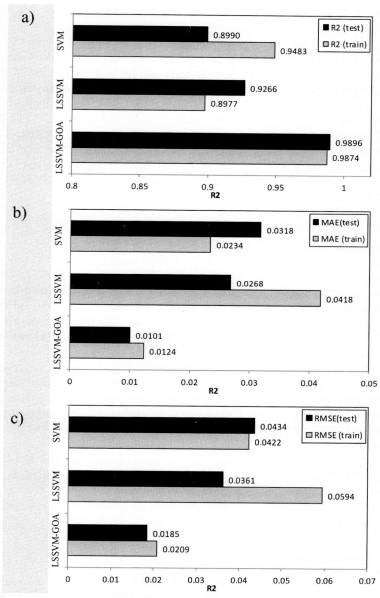

FIG. 7 Comparison of models' accuracy and performance in estimating the flood volume reduction during training and testing phases: (A) R^2, (B) MAE, and (C) RMSE.

Fig. 8 shows the DR value distribution for the LSSSVM-GOA, LSSVM, and SVM models in the training stage. For example, the DR_{max}, DR_{min}, and DR_{ave} values for the LSSSVM-GOA model are 1.2207, 0.8529, and 1.0048, respectively. However, the DR_{max} values for the LSSVM and SVM models are estimated at 1.4328 and 1.3518, respectively. Also, for LSSVM and SVM models, DR_{min} is 0.7705 and 0.7736, respectively. It should be noted that DR_{ave} for LSSVM and SVM models is 1.0157 and 1.0085, respectively.

Fig. 9 shows the distribution of DR values for LSSSVM-GOA, LSSVM, and SVM models in the test period. During this period, the values 1.0155, 1.0734, and 1.1024 for DR_{max} were obtained for LSSVM-GOA, LSSVM, and SVM models, respectively. DR_{min} values of 0.8800, 0.8735, and 0.8431 and DR_{ave} values of 0.9878, 0.9818, and 0.9801 were obtained for LSSVM-GOA, LSSVM, and SVM models, respectively.

3.2.2   Accuracy regarding volume reduction estimates

Fig. 10 shows the box diagrams of volume reduction (VR) results generated in the training and testing stages using SVM, LSSVM, and LSSVM-GOA models. As can be seen, in both the training and testing periods, the LSSVM-GOA model can better simulate VR than the other algorithms, and its distribution is more like the observational data distribution than in the other models (75th and 25th percentiles are approximately the same for observational model and LSSVM-GOA model). This indicates the superiority of this method over other models in predicting VR in both training and testing.

Fig. 11 compares volume reduction (VR) in both observational and predictive modes of the LSSVM-GOA, LSSVM, and SVM models in the training course. This figure also presents the correlation between the real and estimated data. In the training phase, the LSSVM-GOA algorithm also outperformed both SVM and LSSMV based on the highest correlation (0.9874, 0.9483, and 0.8977, respectively).

Fig. 12 examines the correlation between observational data and the output of LSSVM-GOA, LSSVM, and SVM models during the testing period. In this period, the LSSVM-GOA model with a correlation coefficient of 0.9896 has shown better performance than the LSSVM with a value of 0.9266 and the SVM model with a value of 0.8990.

3.3   Reliability of the models

Fig. 13 shows the Taylor diagrams of LSSVM-GOA, LSSVM, and SVM for both train and testing stages. Similar to previous findings, the LSSVM-GOA has the highest accuracy, whereas the SVM model has the lowest accuracy among the studied models in both the training and the testing stages.

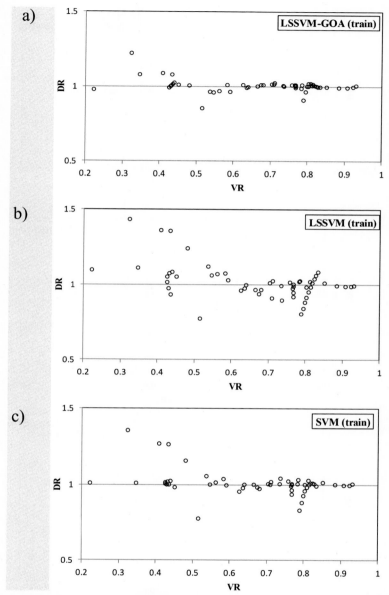

FIG. 8 Dimensionality reduction (DR) graphs in the training phase of the models: (A) LSSVM-GOA, (B) LSSVM, and (C) SVM.

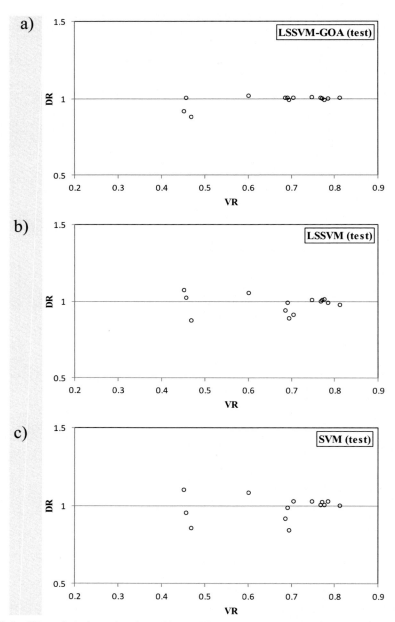

FIG. 9 DR graphs in the testing phase of the model development: (A) LSSVM-GOA, (B) LSSVM, and (C) SVM.

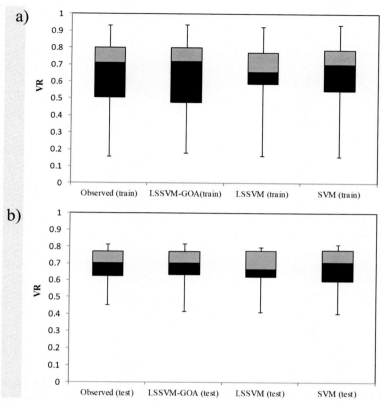

FIG. 10   Boxplots of observed volume reduction compared with predicted volume reduction by LSSVM-GOA, LSSVM, and SVM models in (A) training phase and (B) testing phase.

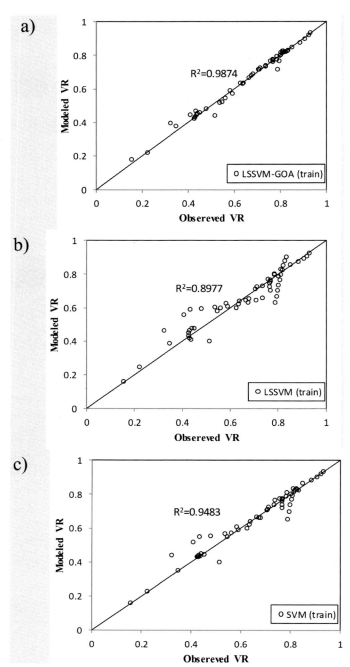

FIG. 11 Correlation between real and estimated data in the training phase: (A) LSSVM-GOA, (B) LSSVM, and (C) SVM.

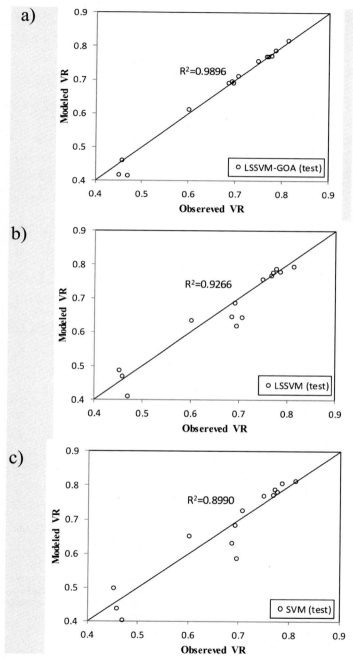

FIG. 12 The correlation of real and estimated data in the testing phase: (A) LSSVM-GOA, (B) LSSVM, and (C) SVM.

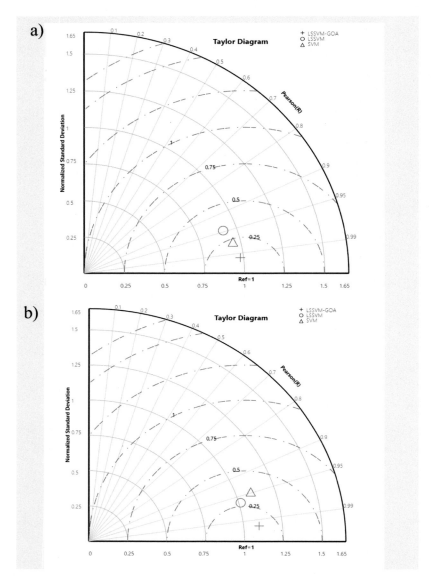

FIG. 13 Taylor diagrams of intelligent algorithm predictions: (A) training phase and (B) testing phase. The longitudinal distance from the origin of the coordinates indicates the standard deviation, the radial lines represent the correlation coefficient, and the linear lines show the RMSE. As the circumference of the circle increases, the value of the mentioned parameter increases.

4 Conclusions

This study investigates the impact of using three types of low-impact development (LID) solutions, i.e., rain barrel (RB), permeable pavement (PP), and infiltration trench (IT). Seven scenarios involving the use of LID as isolated or combined solutions, with 10 plans in each scenario to represent different percentage cover within the Golestan town study site. Based on SWMM modeling, the PP-RB scenario recorded the highest average reduction in flood volume (80%) and peak discharge (90%) compared to the current system. Opposing, the IT scenario showed the weakest performance among the seven scenarios, with an average reduction in peak discharge and total flood volume by up 60% and 40%, respectively.

Calculating the reduced volume of floods during rainfall events is one of the goals of designers of urban runoff control systems. Finding the reduced flood rate always requires long calculations and time-consuming modeling, which involve high costs. The method proposed in this chapter to quickly compare the effectiveness of LIDs in reducing floods is based on using intelligent algorithms by considering only the area of the desired LIDs along with the rate of peak discharge reduction. The intelligent algorithms revealed a good performance in estimating flood reduction. The results obtained from the simulations performed by these algorithms indicate that the LSSVM-GOA hybrid algorithm displayed the highest performance in simulating the flood volume reduction. The high correlation of all three algorithms indicates the high power of all three in predicting the reduced flood volume, while in the training period, the data related to the LSSVM-GOA model have the highest correlation and density among the studied models. These results show that the LSSVM-GOA hybrid algorithm has a good ability to predict the rate of flood reduction. In other words, the proposed method in this research has much less limitations in terms of time and cost of modeling compared to conventional modeling methods such as the SWMM method, and the results obtained are also suitable for using designs related to control systems of urban flooding.

Acknowledgments

Carla Ferreira thanks the National funding by FCT—Foundation for Science and Technology, P.I., through the institutional scientific employment program contract (CEECINST/00077/2021).

References

Aghelpour, P., Kisi, O., & Varahvian, V. (2021). Multivariate drought forecasting in short- and long-term horizons using MSPI and data-driven approaches. *Journal of Hydrologic Engineering, 26*(4). https://doi.org/10.1061/(ASCE)HE.1943-5584.0002059.

Ahiablame, L. M., Engel, B. A., & Chaubey, I. (2012). Effectiveness of low impact development practices: Literature review and suggestions for future research. *Water, Air, and Soil Pollution, 223*(7), 4253–4273. https://doi.org/10.1007/s11270-012-1189-2.

Anandhi, A., Srinivas, V. V., Nanjundiah, R. S., & Kumar, D. N. (2008). Downscaling precipitation to river basin in India for IPCC SRES scenarios using support vector machine. *International Journal of Climatology, 28*(3), 401–420.

Azad, A., Farzin, S., Sanikhani, H., Karami, H., Kisi, O., & Singh, V. P. (2021). Approaches for optimizing the performance of adaptive neuro-fuzzy inference system and least-squares support vector machine in precipitation modeling. *Journal of Hydrologic Engineering, 26*(4). https://doi.org/10.1061/(ASCE)HE.1943-5584.0002069.

Babaei, S., Ghazavi, R., & Erfanian, M. (2018). Urban flood simulation and prioritization of critical urban sub-catchment using SWMM model and PROMETHEE II approach. *Physics and Chemistry of the Earth, 105*(2018), 3–11. https://doi.org/10.1016/j.pce.2018.02.002.

Bisht, D. S., Chatterjee, C., Kalakoti, S., Upadhyay, P., Sahoo, M., & Panda, A. (2016). Modeling urban floods and drainage using SWMM and MIKE URBAN: A case study. *Natural Hazards, 84*, 749–776. https://doi.org/10.1007/s11069-016-2455-1.

Bond, N., Costelloe, J., King, A., Warfe, D., Reich, P., & Balcombe, S. (2014). Ecological risks and opportunities from engineered artificial flooding as a means of achieving environmental flow objectives. *Frontiers in Ecology and the Environment, 12*(7), 386–394. https://doi.org/10.1890/130259.

Choubin, B., Moradi, E., Golshan, M., Adamowski, J., Sajedi-Hosseini, F., & Mosavi, A. (2018). An ensemble prediction of flood susceptibility using multivariate discriminant analysis, classification and regression trees, and support vector machines. *Science of the Total Environment, 651*, 2087–2096. https://doi.org/10.1016/j.scitotenv.2018.10.064.

Dadrasajirlou, Y., Karami, H., & Mirjalili, S. (2023). Using AHP-PROMOTHEE for selection of best low-impact development designs for urban flood mitigation. *Water Resources Management, 37*(1), 375–402.

Dietz, M. E., & Clausen, J. C. (2008). Stormwater runoff and export changes with development in a traditional and low impact subdivision. *Journal of Environmental Management, 87*(4), 560–566. https://doi.org/10.1016/j.jenvman.2007.03.026.

Eckart, J., Sieker, H., Vairavamoorthy, K., & Alsharif, K. (2012). Flexible design of urban drainage systems: Demand led research for Hamburg-Wilhelmsburg. *Reviews in Environmental Science and Biotechnology, 11*(1), 5–10. https://doi.org/10.1007/s11157-011-9256-5.

Faye, M., Traore, V. B., & Bob, M. (2016). Hydrologic modeling in diass river basin using rainfall. *International Journal of Recent Scientific Research, 6*, 7290–7295.

Ferreira, C. S. S., Ferreira, A. J. D., Pato, R. L., Magalhães, M. C., Coelho, C. O., & Santos, C. (2012). Rainfall-runoff-erosion relationships study for different land uses, in a sub-urban area. *Zeitschrift für Geomorphologie, 56*(X), 005–020.

Ferreira, C. S. S., Mourato, S., Ksanin-Grubin, M., Ferreira, A. J. D., Destouni, G., & Kalantari, Z. (2020). Effectiveness of nature-based solutions in mitigating flood hazard in a Mediterranean PeriUrban catchment. *Water, 12*, 2893.

Ferreira, C. S. S., Walsh, R. P. D., Nunes, J. P. C., Steenhuis, T. S., Nunes, M., de Lima, J. L. M. P., Coelho, C. O. A., & Ferreira, A. J. D. (2016). Impact of urban development on streamflow regime of a Portuguese peri-urban Mediterranean catchment. *Journal of Soils and Sediments, 16*, 2580–2593.

Fiori, A., & Volpi, E. (2020). On the effectiveness of LID infrastructures for the attenuation of urban flooding at the catchment scale. *Water Resources Research, 56*(5), e2020WR027121. https://doi.org/10.1029/2020WR027121.

Flood, J. (1997). Urban and housing indicators. *Urban Studies, 34*, 1635–1665. https://doi.org/10.1080/0042098975385.

Freer, J., Beven, J., Schumann, G., Hall, J., & Bates, P. (2013). *Flood risk and uncertainty*. UK: Cambridge University Press.

Fu, X., Hopton, M. E., Wang, X., Goddard, H., & Liu, H. (2019). A runoff trading system to meet watershed-level stormwater reduction goals with parcel-level green infrastructure installation. *Science of the Total Environment, 689*, 1149–1159. https://doi.org/10.1016/j.scitotenv.2019.06.439.

Geng, R., & Sharpley, A. N. (2019). A novel spatial optimization model for achieve the trade-offs placement of best management practices for agricultural non-point source pollution control at multi-spatial scales. *Clean Production, 234*, 1023–1032. https://doi.org/10.1016/j.jclepro.2019.06.277.

Ghazavi, R., Vali, A., & Eslamian, S. (2012). Impact of flood spreading on groundwater level variation and groundwater quality in an arid environment. *Water Resources Management, 26*, 1651–1663. https://doi.org/10.1007/s11269-012-9977-4.

Henonin, J., Hongtao, M., Zheng-Yu, Y., Hartnack, J., Havnø, K., Gourbesville, P., & Mark, O. (2015). Citywide multi-grid urban flood modelling: The July 2012 flood in Beijing. *Urban Water Journal, 12*(1), 52–66. https://doi.org/10.1080/1573062X.2013.851710.

Hoang, L., & Fenner, R. A. (2016). System interactions of stormwater management using sustainable urban drainage systems and green infrastructure. *Urban Water Journal, 13*(7), 739–758. https://doi.org/10.1080/1573062X.2015.1036083.

Huang, C.-L., Hsu, N.-S., Liu, H.-J., & Huang, Y.-H. (2018). Optimization of low impact development layout design for megacity flood mitigation. *Journal of Hydrology, 564*, 542–558. https://doi.org/10.1016/j.jhydrol.2018.07.044.

Jacobson, C. R. (2011). Identification and quantification of the hydrological impacts of imperviousness in urban catchments: A review. *Journal of Environmental Management, 92*(6), 1438–1448. https://doi.org/10.1016/j.jenvman.2011.01.018.

Jia, H., Yao, H., Tang, Y., Yu, S. L., Zhen, J. X., & Lu, Y. (2013). Development of a multi-criteria index ranking system for urban runoff best management practices (BMPs) selection. *Environmental Monitoring and Assessment, 185*(2013), 7915–7933. https://doi.org/10.1007/s10661-013-3144-0.

Karami, H., DadrasAjirlou, Y., Jun, C., Bateni, S. M., Band, S. S., Mosavi, A., Moslehpour, M., & Chau, K.-W. (2022). A novel approach for estimation of sediment load in dam reservoir with hybrid intelligent algorithms. *Frontiers of Environmental Science, 10*, 821079. https://doi.org/10.3389/fenvs.2022.821079.

Kayhanian, M., Fruchtman, B. D., Gulliver, J. S., Montanaro, C., Ranieri, E., & Wuertz, S. (2012). Review of highway runoff characteristics: Comparative analysis and universal implications. *Water Resources, 46*(20), 6609–6624. https://doi.org/10.1016/j.watres.2012.07.026.

Lawson, E., Thorne, C., Ahilan, S., Allen, D., Arthur, S., Everett, G., Fenner, R., Glenis, V., Guan, D., & Hoang, L. (2014). *Delivering and evaluating the multiple flood risk benefits in blue-green cities: An interdisciplinary approach.* WIT Press. https://scholar.google.com/scholar_lookup?title=Delivering%20and%20Evaluating%20the%20Multiple%20Flood%20Risk%20Benefits%20in%20Blue-Green%20Cities%3A%20an%20Interdisciplinary%20Approach&author=E.%20Lawson&publication_year=2014.

Lei, X., Chen, W., Panahi, M., Falah, F., Rahmati, O., Uuemaa, E., Kalantari, Z., Ferreira, C., Rezaie, F., Tiefenbacher, J. P., Lee, S., & Bian, H. (2021). Urban flood modelling using deep-learning approaches in Seoul, South Korea. *Journal of Hydrology, 601*, 126684.

Li, L., Uyttenhove, P., & Vaneetvelde, V. (2020). Planning green infrastructure to mitigate urban surface water flooding risk e a methodology to identify priority areas applied in the city of Ghent. *Landscape and Urban Planning, 194*, 103703. https://doi.org/10.1016/j.landurbplan.2019.103703.

Liu, Y., Ji, Y., Liu, D., Fu, Q., Li, T., Hou, R., Li, Q., Cui, S., & Li, M. (2021). A new method for runoff prediction error correction based on LS-SVM and a 4D copula joint distribution. *Journal of Hydrology, 598*, 126223. https://doi.org/10.1016/j.jhydrol.2021.126223.

Mark, O., Weesakul, S., Apirumanekul, C., Aroonnet, S. B., & Djordjevic, S. (2004). Potential and limitations of 1D modelling of urban flooding. *Journal of Hydrology*, *299*(3–4), 284–299. https://doi.org/10.1016/j.jhydrol.2004.08.014.

Martin-Mikle, C. J., Beurs, K. M., Julian, J. P., & Mayer, P. M. (2015). Identifying priority sites for low impact development (LID) in a mixed-use watershed. *Landscape and Urban Planning*, *140*, 29–41. https://doi.org/10.1016/j.landurbplan.2015.04.002.

Miguez, M. G., Battemarco, B. P., De Sousa, M. M., Rezende, O. M., Verol, A. P., & Gusmaroli, G. (2017). Urban flood simulation using MODCEL—An alternative quasi2D conceptual model. *Water*, *9*(6), 445. https://doi.org/10.3390/w9060445.

Olang, L. O., & Furst, J. (2011). Effects of land cover change on flood peak discharges and runoff volumes: Model estimates for the Nyando River Basin, Kenya. *Hydrological Processes*, *25*(1), 80–89. https://doi.org/10.1002/hyp.7821.

Palla, A., & Gnecco, I. (2015). Hydrologic modeling of low impact development systems at the urban catchment scale. *Journal of Hydrology*, *528*, 361–368. https://doi.org/10.1016/j.jhydrol.2015.06.050.

Panahi, M., Rahmati, O., Kalantari, Z., Darabi, H., Rezaie, F., Moghaddam, D. D., Ferreira, C. S. S., Foody, G., Lee, C. W., & Lee, S. (2022). Large-scale dynamic flood monitoring in an arid-zone floodplain using SAR data and hybrid machine learning models. *Journal of Hydrology*, *611*, 128001.

Pappalardo, V., La Rosa, D., Campisano, A., & La Greca, P. (2017). The potential of green infrastructure application in urban runoff control for land use planning: A preliminary evaluation from a southern Italy case study. *Ecosystem Services*, *26*, 345–354. https://doi.org/10.1016/j.ecoser.2017.04.015.

Park, D., Song, Y. I., & Roesner, L. A. (2013). Effect of the seasonal rainfall distribution on stormwater quality capture volume estimation. *Journal of Water Resources Planning and Management*, *139*(1), 45–52. https://doi.org/10.1061/(ASCE)WR.1943-5452.0000204.

Pham, Q. B., Yang, T. C., Kou, C. M., Tseng, H. W., & Yu, P. S. (2021). Coupling singular spectrum analysis with least square support vector machine to improve accuracy of SPI drought forecasting. *Water Resources Management: An International Journal*, *35*(3), 847–868. https://doi.org/10.1007/s11269-020-02746-7.

Randhir, T. O., & Raposa, S. (2014). Urbanization and watershed sustainability: Collaborative simulation modeling of future development states. *Journal of Hydrology*, *519*(part B), 1526–1536. https://doi.org/10.1016/j.jhydrol.2014.08.051.

Razavi, S., Gober, P., Maier, H. R., Brouwer, R., & Wheater, H. (2020). Anthropocene flooding: Challenges for science and society. *Hydrological Processes*, *34*(8), 1996–2000. https://doi.org/10.1002/hyp.13723.

Rene, J. R., Djordjevic, S., Butler, D., Madsen, H., & Mark, O. (2014). Assessing the potential for real-time urban flood forecasting based on a worldwide survey on data availability. *Urban Water Journal*, *11*(7), 573–583. https://doi.org/10.1080/1573062X.2013.795237.

Saremi, S., Mirjalili, S., & Lewis, A. (2017). Grasshopper optimisation algorithm: Theory and application. *Advanced Software Engineering*, *105*, 30–47.

Schoonover, J. E., Lockaby, B. G., & Helms, B. S. (2006). Impacts of land cover on stream hydrology in the West Georgia piedmont, USA. *Journal of Environmental Quality*, *35*(6), 2123–2131. https://doi.org/10.2134/jeq2006.0113.

Shahed Behrouz, M., Zhu, Z., Matott, L. S., & Rabideau, A. J. (2020). A new tool for automatic calibration of the Storm water Management Model (SWMM). *Journal of Hydrology*, *581*, 124436. https://doi.org/10.1016/j.jhydrol.2019.124436.

Shen, Z., Hou, X., Li, W., & Aini, G. (2014). Relating landscape characteristics to nonpoint source pollution in a typical urbanized watershed in the municipality of Beijing. *Landscape and Urban Planning, 123*(96), 107. https://doi.org/10.1016/j.landurbplan.2013.12.007.

Song, C., Yao, L., Hua, C., & Ni, Q. (2021). A water quality prediction model based on variational mode decomposition and the least squares support vector machine optimized by the sparrow search algorithm (VMD-SSA-LSSVM) of the Yangtze River., China. *Environmental Monitoring and Assessment, 193*(6), 363. https://doi.org/10.1007/s10661-021-09127-6.

Suykens, J. A. (2001). Nonlinear modelling and support vector machines. In *Vol. 1. IMTC 2001, proceedings of the 18th IEEE Instrumentation and measurement technology conference* (pp. 287–294).

Tavakol-Davani, H., Gohraian, E., Hansen, C. H., Tavakol-Davani, H., Apul, D., & Burian, S. J. (2016). How does climate change affect combined swer overflow in a system benefiting from rainwater harvesting systems? *Sustainable Cities and Society, 27*(2016), 430–438. https://doi.org/10.1016/j.scs.2016.07.003.

Teng, J., Jakeman, A., Vaze, J., Croke, B., Dutta, D., & Kim, S. (2017). Flood inundation modelling: A review of methods, recent advances and uncertainty analysis. *Environmental Modelling & Software, 90*, 201–216. https://doi.org/10.1016/j.envsoft.2017.01.006.

Teymouri, E., Mousavi, S.-F., Karami, H., Farzin, S., & Kheirabad, H. (2020). Reducing urban runoff pollution using porous concrete containing mineral adsorbents. *Journal of Environmental Treatment Techniques, 8*(1), 429–436. http://www.jett.dormaj.com/docs/Volume8/Issue%201/Reducing%20Urban%20Runoff%20Pollution%20Using%20Porous%20Concrete%20Containing%20Mineral%20Adsorbents.pdf.

UN. (2012). *World urbanization prospects: The 2011 revision: United Nations.* Department of Economic and Social Affairs. Population Division. https://www.un.org/en/development/desa/population/publications/pdf/urbanization/WUP2011_Report.pdf.

Vapnik, V. N. (1995). *The nature of statistical learning theory.* New York: Springer.

Wang, Y., Choi, W., & Deal, B. M. (2005). Long-term impacts of land-use change on non-point source pollutant loads for the St. Louis metropolitan area, USA. *Environmental Management, 35*, 194–205. https://doi.org/10.1007/s00267-003-0315-8.

Wolch, J. R., Byrne, J., & Newell, J. P. (2014). Urban green space, public health, and environmental justice: The challenge of making cities 'just green enough'. *Landscape and Urban Planning, 125*, 234–244. https://doi.org/10.1016/j.landurbplan.2014.01.017.

Zhang, W., Villarini, G., Vecchi, G. A., & Smith, J. A. (2018). Urbanization exacerbated the rainfall and flooding caused by hurricane Harvey in Houston. *Nature, 563*(7731), 384–388. https://doi.org/10.1038/s41586-018-0676-z.

Zhang, B., Xie, G., Li, N., & Wang, S. (2015). Effect of urban green space changes on the role of rainwater runoff reduction in Beijing, China. *Landscape and Urban Planning, 140*, 8–16. https://doi.org/10.1016/j.landurbplan.2015.03.014.

Zu, M., Chen, w., Hu, Y., Hou, Y., Liu, L., & Zhang, K. (2021). DRGraph: An efficient graph layout algorithm for large-scale graphs by dimensionality reduction. *IEEE Transactions on Visualization and Computer Graphics, 27*(2), 1666–1676.

Chapter 1.5

Policy nexus for urban resilience and nature-based solutions

Nina Escriva Fernandez[a] and Haozhi Pan[b]

aSchool of International and Public Affairs, Shanghai Jiao Tong University, Shanghai, China,
bShanghai Jiao Tong University, Shanghai, China

1 Introduction

Cities do not only have the potential to make a change in terms of climate change but are also especially at risk for its negative consequences, such as flooding and heat stress, with potentially relevant social and economic effects (Revi et al., 2014; Tol, 2018; Wang, 2022). Resilience refers to the capacity of a city to prepare for future shocks and stresses and the ability to handle crises as they arrive (Coaffee et al., 2018). Policy aiming to increase urban resilience is seen as an effective way to address issues of climate change and the potential social and economic consequences that can come from it.

In response to these challenges, numerous multilateral organizations have shifted their focus toward resilience, initiating programs aimed at bolstering the resilience of cities. One such initiative is UN Habitat's City Resilience Global Programme (UN Habitat, 2020), orchestrated by the Urban Resilience Hub. This program is designed to strengthen municipal capacities in addressing urban resilience-related obstacles. Through the provision of vulnerability and risk assessments (UN Habitat, 2020, 2021), the organization extends its support to governments worldwide. The consensus within multilateral frameworks targeting resilience underscores the positive correlation between urban resilience and the mitigation of climate change effects. Urban Resilience Hub, acting as a technical partner of the UN Habitat, characterizes urban resilience as the cornerstone for a sustainable urban future. Consequently, numerous cities globally have pledged to embrace the concept of resilience, aligning their policies with the initiatives set forth by multilateral organizations (Coaffee et al., 2018).

Moreover, nature-based solutions (NbS) have emerged as a pivotal component of the urban resilience agenda, witnessing a surge in pilot projects and attracting substantial investments (Zari et al., 2019). The preservation and fostering of urban ecosystem services, coupled with the transition from

Nature-Based Solutions in Supporting Sustainable Development Goals
https://doi.org/10.1016/B978-0-443-21782-1.00011-7

conventional gray infrastructure to greener alternatives, are indispensable for fostering resilient ecosystems and promoting societal well-being (Lafortezza et al., 2018). This imperative is particularly pronounced in the Global South, where the boundaries between urban and rural areas are often blurred.

Governments, financial institutions, and international agencies are fervently advocating for NbS due to their potential to address a myriad of interconnected challenges spanning climate, biodiversity, and social domains (Girardin et al., 2021; Goodwin et al., 2023; Seddon et al., 2020, 2021). NbS projects are envisioned to mitigate the adverse impacts of climate change while augmenting the quantity and quality of ecosystem services. Concurrently, resilience encompasses not only the ability of urban ecosystems to rebound but also their contribution to the resilience of infrastructure and social systems (Folke, 2016). Thus, the resilience discourse captures the multifaceted interplay between urban ecosystems and the broader fabric of urban resilience.

On the European level, both resilience and NbS have also been identified as a priority. Recently, all member states of the European Union were required to submit a resilience plan in response to the consequences of COVID-19 (European Commission, 2022a, 2022b). In October 2022, the Dutch government published its National Resilience Plan through the Ministry of Finance (Ministerie van Financiën, 2022). Several years before this national plan was written and approved by the EU; however, many cities in the Netherlands had already adopted the idea of resilience into their policy, and several cities had joined international resilience networks; examples are Rotterdam and The Hague in the Rockefeller Foundation 100 Resilient Cities Network (Rockefeller Foundation, 2020). On the other contrary, at the current moment, many Dutch cities have still not integrated the concept of resilience into their policy. In light of this, the chapter will start with understanding the policy nexus between NbS and urban resilience by analyzing existing literature and cases and then surveying the existing landscapes of municipalities in the Netherlands that adopt the resilience policy. Our main research questions are: (1) What are policy nexus, barriers, and potentials for urban resilience and NbS? (2) Have the largest 50 municipalities in the Netherlands integrated the concept of resilience into their policy?

Research on the number of cities that have integrated the concept of resilience into policy is a starting point to gain more insight into the drivers behind the adoption of resilience, NbS, and other climate-focused policies. Which can then lead to better strategies to incentivize the adoption of said policies.

2 Urban resilience

In 2016, at the Habitat III global conference, the United Nations adopted the New Urban Agenda. This agenda puts great emphasis on building resilience and emphasizes the role of cities in reducing greenhouse gas emissions (United Nations, 2016). The UN Sustainable Development Goals also place a focus on resilience. The Sustainable Development Goal #11, Sustainable

Cities and Communities, that reads "Make cities and human settlements inclusive, safe, resilient and sustainable" (United Nations Statistics Division, 2019) is most associated with resilience. However, resilience is also mentioned in Sustainable Development Goals 1, 2, 3, 9, 13, and 14. UN Habitat also refers to Sustainable Development Goals 7, 9, 23, 29, and 33 in relation to resilience, which is indicative of the multidisciplinary character of the resilience concept (Urban Resilience Hub, 2018). The concept of resilience can sometimes describe the aim to create resilience in public management (Steccolini et al., 2017), but the resilience frameworks utilized and developed by international actors such as the UN generally target urban or community resilience.

The academic debate about the term resilience has been in process for several decades and continues to be ongoing (Eraydin & Özatağan, 2021; Lu & Stead, 2013; Meerow & Newell, 2019; Torabi et al., 2021). According to some scholars, the concept is theoretical and difficult to grasp but also provides flexibility in implementation (Sharma et al., 2023). However, the resilience frameworks as developed by multilateral actors such as the UN and World Bank are similar in nature. These frameworks have been adopted by governments worldwide in the policy-making process and include different phrasings of the concept "resilient city." Nevertheless, they all refer to the capacity of a city to prepare for future shocks and stresses, to both adapt and to cope (Sharma et al., 2023), and the ability to handle crises as they arrive (Coaffee et al., 2018). This is a shift away from a common prior singular focus on vulnerabilities (Béné et al., 2015). The OECD defines a resilient city as "cities that have the ability to absorb, recover and prepare for future shocks." They identify four drivers of resilience: economy, government, society, and environment.

Global cities such as London and New York have taken the concept of resilience and adopted it to their own needs (Wang, 2021). This is in part because the risks and stresses faced, and the potential social and economic consequences of these, are dependent on the specific city (Naef, 2022; Tol, 2018; Wang, 2021, 2022). Resilience can thus be seen as a "context-specific notion" (Beauchamp et al., 2019). In addition to that, policy addressing resilience is often a "policy mix": various goals and instruments addressing the issue (Lesnikowski et al., 2019). In practice, this means that the interpretation of resilience and the implemented policies in order to achieve resilience is up to interpretation by the individual government units. For that reason, this article will focus on municipalities in the Netherlands integrating the concept of resilience into their policy strategies without looking at the specific policies that will be implemented to reach resilience.

Utilizing the various known definitions of resilience, the United Nation's interdisciplinary approach to resilience, and the OECD's four drivers of resilience (Sharma et al., 2023; Urban Resilience Hub, 2018), this research project will use the following definition of resilience: the ability to prepare for and handle future shocks and stresses on the economy, government, society, and environment. As this research project focuses on the integration of the concept of

TABLE 1 Municipality classification criteria.

Classification number	Classification criteria
Classification 1	No resilience plans or integration of the concept of resilience in strategic policy documents in either English or Dutch
Classification 2	No resilience plans in either English or Dutch, BUT the concept of resilience is integrated into strategic policy documents
Classification 3	One or more resilience plans focusing on one sector of resilience (environment, society, economy, government) in Dutch, AND the concept of resilience is integrated into strategic policy documents
Classification 4	Overarching resilience plan focusing on at least three sectors of resilience in English

resilience into policy but not on the specific policies, resilience policy will be defined as follows: a strategy that acknowledges becoming resilient as something beneficial and seeks for the unit of government to become resilient (able to prepare for and handle future shocks and stresses on the economy, government, society, and environment) (see Table 1).

3   The nexus between NbS and urban resilience

NbS can address various climate hazards, such as intense precipitation, rising temperatures, droughts, and coastal hazards. NbS also improves resilience in the fields of coastal management, securing resources, and diminishing socioeconomic vulnerability to a number of risks. NbS interventions encompass a range of measures, including reducing exposure to climatic hazards, restoring natural habitats, integrating natural and built infrastructure, and empowering citizens to enact their own adaptation strategies (Goodwin et al., 2023; Zari et al., 2019). NbS also alleviates interlinked pressures stemming from environmental degradation, such as water pollution, soil sealing due to urban expansion, deforestation from illegal logging, and biodiversity loss.

For example, urban green spaces, as a type of NbS, play a crucial role in bolstering resilience against the chronic stresses and gradual changes to which cities are exposed (Bush & Doyon, 2019). A combination of NbS could involve implementing agroforestry systems to restore wetlands for local food supply or flood risk mitigation, alongside mangrove restoration (Kalantari et al., 2018). The hydrological benefits of vegetation, such as stabilizing slopes to prevent rainfall-induced landslides, are well-documented (Gonzalez-Ollauri & Mickovski, 2017). Additionally, the effectiveness of NbS in stormwater management has been extensively researched (Beceiro et al., 2022; Staddon et al., 2018). Moreover, discussions regarding the nexus between resilience and NbS should encompass the resilience of ecosystems themselves (Bush & Doyon, 2019). Ecosystem resilience necessitates dynamic urban systems capable of adapting to the impacts of climate change, environmental shifts, and the pressures of urbanization.

NbS offer a distinctive advantage in spatial distribution. They function as decentralized, distributed systems for infrastructure service provision, inherently more resilient than large centralized gray infrastructure (Bush & Doyon, 2019). Utilizing future scenario planning and predictive GIS techniques is crucial for NbS to proactively address potential challenges rather than merely reacting to current issues. This proactive approach is particularly vital in the Global South due to governance challenges, rapid population growth, high densities, ocean and weather complexities, food and water security concerns, inadequate land use planning, and pollution management (Zari et al., 2019).

As examples of spatial mapping and allocation for NbS, Goodwin et al. (2023) extensively map and analyze NbS approaches tailored for climate change adaptation worldwide. They highlight limitations in current NbS practices, particularly in addressing multidimensional climate vulnerability, social justice, fostering public-private collaborations, and realizing co-benefits. Additionally, they underscore the bias toward knowledge and practice favoring the Global North, emphasizing the need to harness the transformative potential of urban NbS. Moreover, exploring innovative applications of existing ecosystems is imperative. For example, consider the role of urban woodlands and street trees in climate change adaptation or the contribution of urban parks to social cohesion (Maes et al., 2016; Pan et al., 2023). Beceiro et al. (2022) advocate for a comprehensive Resilience Assessment Framework (RAF) to evaluate NbS contributions to urban resilience, with a focus on stormwater management and control, as demonstrated in a case study application.

The benefits of NbS are often intricate and interconnected. NbS has largely evolved from previous concepts and principles, including sustainability, resilience, and ecosystem services, with a central emphasis on generating multiple co-benefits for the environment, economy, and society (Davies & Lafortezza, 2017). According to the Green Infrastructure Strategy report by the European Commission (EC), NbS can offer less expensive, cost-effective, and enduring solutions (Kabisch et al., 2016; Lafortezza et al., 2018). When carefully and comprehensively designed, NbS can foster win-win situations across interconnected social, ecological, and economic systems. It is crucial to understand co-benefits holistically to avoid diminishing or trading off multiple ecosystem services (Zari et al., 2019). Zari et al. (2019) discuss the potential of NbS in Pacific Small Island Developing States as a response to climate change. They stress the importance of addressing interlinked ecological, climate, and human well-being issues along with associated co-benefits. Elmqvist et al. (2015) assert that investing in restoring green and blue infrastructure in urban areas may not only be ecologically and socially desirable but also economically advantageous.

4 Policy actions for NbS and urban resilience

Policy integrations, implementation, and eventually scaling-up of NbS and urban reliance are pivotal in addressing multifaceted challenges. However, the implementation of resilient policies often encounters complexities and criticisms. On the one hand, there is a huge room for cities worldwide to learn existing policy tools and instruments from each other. On the other hand, NbS and urban resilience policies need to be adapted and complementary to local contexts. This section discusses some of the policy actions, barriers, and issues in integrating NbS and urban resilience.

Local context is a "dual-edge" sword toward NbS and urban resilience policy implementation. Many resilient policies are challenged by indigenous and local community groups, grassroots organizations, and certain nations (especially in the Global South) for downplaying their social justice implications and potential for maladaptive outcomes. On the contrary, the practice of working closely with nature to create effective human set elements while maintaining healthy ecosystems is a cornerstone of many indigenous belief systems (Zari et al., 2019). Thus, NbS should be more commonly framed as acknowledging and complementary to the local context. NbS can be more effective and culturally appropriate manner than other types of policy strategies (Cousins, 2021; Goodwin et al., 2023; Melanidis & Hagerman, 2022; Seddon, 2022; Westman & Castán Broto, 2022; Zari et al., 2019). It is imperative to employ ecosystem services in tandem with innovative approaches without exacerbating local vulnerability (Liquete et al., 2016).

At the same time, the ecosystems management ability should be improved along with new governance forms with distinct accountability processes. Incorporating ecosystems into urban resilience should avoid maladaptive measures. For instance, if the adaptive decisions neglect that urban growth can influence the environmental exposure level, they also ignore retroactive effects for the most vulnerable groups. NbS and urban resilience have received strong policy support, including the Millennium Ecosystem Assessment within the framework of the Sustainable Development Goals and the Strategic Plan for Biodiversity 2011–20 under the Convention on Biological Diversity. Successful adaptation requires an understanding of interactions within socio-ecological systems (Murphy et al., 2016; Young et al., 2019).

5 Urban resilience policies in the Netherland

Policy nexuses for urban resilience and NbS are potentially beneficial and produce additional co-benefits, but they are also faced with many challenges. How is the policy landscape right now? To answer this key research question, we examine whether the largest 50 municipalities (by population) in the Netherlands integrated the concept of resilience into their policy through policy document collection, analysis, and classification.

5.1 Policy document collection

The first stage of data collection consisted of the assortment of different policy documents from each of the 50 municipalities. This was carried out in three different steps: (1) Collection of thematic policy documents on resilience through the municipal website, (2) web search on municipality and resilience, and (3) collection of strategic policy documents part of the municipal policy cycle through the appropriate municipal channels.

(1) The first step consisted of the collection of thematic on resilience through the municipal website. A search was performed on each of the municipalities' websites for the English term "resilience," and a list of Dutch terms that are translations of the word resilience or strongly associated with the word resilience. These terms were searched for in adjective form and noun form. In the Netherlands, there are multiple translations of the word resilience. These terms are often used in reference to a specific type of resilience. For example, "wendbaar" is commonly used in relation to governmental resilience. In addition, to perform a better-quality search, the translation of the word vulnerable was searched for, as resilience policy is often carried out in response to vulnerabilities, but the policy document was only included if it then directly referred to creating resilience as opposed to describing vulnerabilities.

Another point of interest is the interchangeable use of the terms climate adaptive and climate resilient in the Netherlands. Because the reasoning and approach to climate adaptivity and climate resilience are not the same, only the term climate resilience was included in the search. Results based on the term climate adaptive were left out. This is because the scope of this research focuses specifically on resilience and not on climate adaptation policy or general climate policy that does not focus on creating resilience within the municipality.

A total of seven terms (+ their adjective forms) were searched for during this step of the study (Table 2).

(2) The second step consisted of a web search on the municipality and resilience. A search engine search (Google) was using the different Dutch terms used for resilience as listed in Table 2: [resilience + city name] and [resilient + city name]. This step was added to control for the different quality of search engines on the municipal websites, as well as the development of resilience-policy websites that target English-speaking residents of the city and/or international nonresidents.

The results from the searches performed, using the seven resilience-related keywords, included relevant policy documents but also included event announcements, news articles, speeches given by municipal officials, and general interest articles. These were left out of the study. The collected documents in the two steps described earlier had to fulfill a few requirements to be included in the analysis:

TABLE 2 Keywords utilized in policy document collection.

Keyword	Translation
Klimaatbestendig/klimaatbestendigheid/klimaatbestendige	Climate resilient
Kwetsbaar/kwetsbaarheid/kwetsbare	Vulnerable
Resilient/resilience	Resilient
Toekomstbestendig/toekomstbestendigheid/toekomstbestendige	Futureproof
Veerkracht/veerkrachtig/veerkrachtige	Resilient
Weerbaar/weerbaarheid/weerbare	Resilient
Wendbaar/wendbaarheid/wendbare	Resilient

(A) The document must be a thematic policy strategy/a strategic policy document. A thematic policy strategy describes the broader ideas or strategy behind policy measures. This research focuses on the integration of the concept of resilience with the aim of making concrete policy measures as a result of the strategy in the future. What this means concretely is that documents describing a single municipal policy, subsidy, rule, or project were left out. This is because those are activities that are carried out as a result of the strategic policy document. For example, if there is a single policy aiming at creating resilience, a focus on resilience should have also been mentioned in a strategic policy document.

(B) The policy document must have been approved through the official municipal policy-making channel: a vote in the city council or approval during a city council meeting.

Some of the documents came up multiple times using different search terms. These policy documents were included in the study only once.

(3) The third step consisted of the collection of strategic policy documents as part of the municipal policy cycle through the appropriate municipal channels. This step was added in order to collect and analyze policy documents from every municipality and to control the different quality search engines of the municipalities. This ensured that even the municipalities that do not have policy documents mentioning resilience that were collected in steps 1 and 2 would have policy documents to analyze in the study. The third step consisted of the collection of policy documents that are part of every municipality's policy cycle, which generally lasts 4 years. Municipalities in the Netherlands publish certain documents on a cyclical basis to justify their activities and corresponding finances to the city's residents.

The documents that were collected from each municipality were the "begroting," which translates to the budget. This document is published by each municipality on a yearly basis and includes the thematic focuses for the upcoming year and the corresponding finances allocated for each of these priorities. This document was collected for every municipality for each year between the years 2018 and 2022 (five documents per municipality).

The second document that was collected was the "akkoord," an agreement between the different governing parties. It is often referred to as "coalitieakkoord" (coalition agreement) but has different names such as "bestuursakkoord" (governmental agreement), among other names. In the Netherlands, the governing parties need to have a majority of the seats in the city council. Therefore, municipal governments often have different parties ruling together. At the start of the administrative period, after elections, the parties must come together to establish an agreement in which they specify their priorities and efforts for the upcoming 4 years. The most recent agreement was collected for each municipality due to the recent elections mostly dating from 2022.

The documents that were not available on the municipal websites were obtained through direct contact with the city council secretariat.

5.2 Policy document analysis

After the collection of all the relevant policy documents in the first step (Policy Document Collection), the conceptual content analysis was performed. The policy documents were searched for the seven keywords also used to collect the documents. This was done in order to identify the resilience-related content of the document. The content of the document was then manually coded and documented for each city according to the following questions:

For the thematic policy documents:

(A) Can this be defined as resilience policy according to the definition this study utilizes: "A strategy that acknowledges becoming resilient as something beneficial and seeks for the unit of government to become resilient (able to prepare for and handle future shocks and stresses on the economy, government, society, and environment)"?

(B) If yes, what sectors of resilience (economy, government, society, environment) are targeted?

For the strategic policy documents:

(A) Does this document integrate the concept of resilience into the policy and thus "acknowledges becoming resilient as something beneficial and seeks for the unit of government to become resilient (able to prepare for and handle future shocks and stresses on the economy, government, society, and environment)"?

(B) If yes, what sectors of resilience (economy, government, society, environment) are targeted?

It is important to distinguish between "nonsubstantive references" and relevant references when conducting content analysis (De Ruiter & Schalk, 2017). Referral to resilience in nonpolicy-related text, such as the introduction of the document, did not qualify as resilience policy.

5.3   Municipality classification

Classification has been used in policy adoption research to identify cities as leaders, followers, or laggards (Otto et al., 2021). Similarly, according to the findings documented in step 2 (Policy Document Analysis), the municipalities were classified according to four classifications that were created inductively based on preliminary research during the content analysis process.

5.4   Results of the policy document collection, analysis, and classification

In order to find the answer to the first research question, a total of 101 thematic policy documents were collected. In addition to that, another 300 strategic policy documents were included. This comes to a total of 401 policy documents having been subject to content analysis. The majority of the analyzed documents were published solely in Dutch, with the exception of a few cities, such as The Hague and Maastricht, that generally have a large expat population. The data collection shows that not every municipality has adopted a resilience policy. In Table 3, we find, however, that all 50 municipalities have included resilience as a priority in their strategic policy document.

TABLE 3 Municipality classification results.

Classification number	Classification description	Number of municipalities
Classification 1	No resilience plans or integration of the concept of resilience in strategic policy documents in either English or Dutch	0
Classification 2	No resilience plans in either English or Dutch, BUT the concept of resilience is integrated into strategic policy documents	27
Classification 3	One or more resilience plans focusing on one sector of resilience (environment, society, economy, government) in Dutch, AND the concept of resilience is integrated into strategic policy documents	20
Classification 4	Overarching resilience plan focusing on at least three sectors of resilience in English	3

Twenty-seven municipalities only mention resilience as a priority in their strategic policy documents. In all these municipalities, resilience is not the main focus of these documents. Twenty municipalities have adopted a resilience policy per the definition of this research article ("A strategy that acknowledges becoming resilient as something beneficial and seeks for the unit of government to become resilient ("able to prepare for and handle future shocks and stresses on the economy, government, society, and environment") (Coaffee et al., 2018; OECD, n.d.-a, n.d.-b; Sharma et al., 2023). All these municipalities have also mentioned resilience as a priority in their strategic policy documents. Nevertheless, only three municipalities have developed an overarching resilience plan focusing on at least three aspects of resilience. The findings are presented in Table 3.

The fact that all municipalities have adopted resilience policies, but only three municipalities have done so in the interdisciplinary way in which resilience is commonly referred to, can point to the difficulties that municipalities experience having to work cross-sectoral (Coaffee et al., 2018; Huck et al., 2020; Lee, 2013). It is possible that individual departments acknowledge the importance of resilience and make a decision to adopt resilience into their policy without having to work with other departments to make the policy cross-sectoral. Since individuals can have a large impact on the policy process, it is possible that the influence of an individual in support of resilience only extends to their own department (Blatter et al., 2021; Pereira, 2022; Yi & Liu, 2022). In addition to that, other potential reasons for the lack of resilience policy are a lack of awareness of the potential benefits of adopting a resilience framework or a lack of knowledge on how to do so.

6 Conclusion

This chapter delves into the intricate policy nexus between nature-based solutions (NbS) and urban resilience, offering a normative examination of their integration derived from pertinent literature. Subsequently, it delves into empirical exploration by analyzing existing policy documents at the municipality level in the Netherlands. In conclusion, the multifaceted benefits of NbS underscore their pivotal role in addressing a myriad of climate hazards and enhancing urban resilience, from mitigating intense precipitation and coastal hazards to bolstering resource security and diminishing socioeconomic vulnerability. Realizing the full potential of NbS requires proactive planning, innovative applications, and comprehensive evaluation frameworks to harness their transformative capacity. By prioritizing co-benefits across social, ecological, and economic domains, NbS not only offer enduring solutions to urban challenges but also foster win-win outcomes for both people and the ecosystem.

For the adoption of urban resilience policy in the Netherlands, it is noteworthy that not every municipality has fully embraced resilience policies, while all municipalities prioritize resilience in their strategic policy documents. At the same time, resilience is mentioned as a priority by most of the cities, though comprehensive resilience planning is lacking. These findings underscore the current state of resilience integration at the municipality level and highlight areas where further policy development is needed to enhance urban resilience across the Netherlands.

References

Beauchamp, E., Abdella, J., Fisher, S., McPeak, J., Patnaik, H., Koulibaly, P., … Gueye, B. (2019). Resilience from the ground up: How are local resilience perceptions and global frameworks aligned? *Disasters*, *43*, 295–317. https://doi.org/10.1111/disa.12342.

Beceiro, P., Brito, R. S., & Galvão, A. (2022). Nature-based solutions for water management: Insights to assess the contribution to urban resilience. *Blue-Green Systems*, *4*(2), 108–134.

Béné, C., Frankenberger, T., & Nelson, S. (2015). Design, monitoring and evaluation of resilience interventions: Conceptual and empirical considerations. In *Institute of Development Studies working paper 459*.

Blatter, J., Portmann, L., & Rausis, F. (2021). Theorizing policy diffusion: From a patchy set of mechanisms to a paradigmatic typology. *Journal of European Public Policy*, *29*(6), 805–825. https://doi.org/10.1080/13501763.2021.1892801.

Bush, J., & Doyon, A. (2019). Building urban resilience with nature-based solutions: How can urban planning contribute? *Cities*, *95*, 102483.

Coaffee, J., Therrien, M. C., Chelleri, L., Henstra, D., Aldrich, D. P., Mitchell, C. L., … Rigaud, É. (2018). Urban resilience implementation: A policy challenge and research agenda for the 21st century. *Journal of Contingencies & Crisis Management*, *26*(3), 403–410. https://doi.org/10.1111/1468-5973.12233.

Cousins, J. J. (2021). Justice in nature-based solutions: Research and pathways. *Ecological Economics*, *180*, 106874.

Davies, C., & Lafortezza, R. (2017). Urban green infrastructure in Europe: Is greenspace planning and policy compliant? *Land Use Policy*, *69*, 93–101.

De Ruiter, R., & Schalk, J. (2017). Explaining cross-national policy diffusion in national parliaments: A longitudinal case study of plenary debates in the Dutch parliament. *Acta Politica*, *52*(2), 133–155. https://doi.org/10.1057/ap.2015.29.

Elmqvist, T., Setälä, H., Handel, S. N., van der Ploeg, S., Aronson, J., Blignaut, J. N., … de Groot, R. (2015). Benefits of restoring ecosystem services in urban areas. *Current Opinion in Environment Sustainability*, *14*, 1010–1108.

Eraydin, A., & Özatağan, G. (2021). Pathways to a resilient future: A review of policy agendas and governance practices in shrinking cities. *Cities*, *115*. https://doi.org/10.1016/j.cities.2021.103226.

European Commission. (2022a). *Recovery and resilience facility*. European Commission. https://ec.europa.eu/info/business-economy-euro/recovery-coronavirus/recovery-and-resilience-facility_en.

European Commission. (2022b). *Recovery and resilience plan for the Netherlands*. European Commission. https://ec.europa.eu/info/business-economy-euro/recovery-coronavirus/recovery-and-resilience-facility/recovery-and-resilience-plan-netherlands_en.

Folke, C. (2016). Resilience (republished). *Ecology and Society*, *21*(4), 21–35. https://doi.org/10.1111/ajps.12521.

Girardin, C. A. J., et al. (2021). Nature-based solutions can help cool the planet—If we act now. *Nature*, *593*, 191–194.

Gonzalez-Ollauri, A., & Mickovski, S. B. (2017). Hydrological effect of vegetation against rainfall-induced landslides. *Journal of Hydrology*, *549*, 374–387.

Goodwin, S., Olazabal, M., Castro, A. J., & Pascual, U. (2023). Global mapping of urban nature-based solutions for climate change adaptation. *Nature Sustainability*, *6*(4), 458–469.

Huck, A., Monstadt, J., & Driessen, P. (2020). Mainstreaming resilience in urban policy making? Insights from Christchurch and Rotterdam. *Geoforum*, *117*, 194–205. https://doi.org/10.1016/j.geoforum.2020.10.001.

Kabisch, N., Frantzeskaki, N., Pauleit, S., Naumann, S., Davis, M., Artmann, M., … Bonn, A. (2016). Nature-based solutions to climate change mitigation and adaptation in urban areas: Perspectives on indicators, knowledge gaps, barriers, and opportunities for action. *Ecology and Society*, *21*(2).

Kalantari, Z., Ferreira, C. S. S., Keesstra, S., & Destouni, G. (2018). Nature-based solutions for flood-drought risk mitigation in vulnerable urbanizing parts of East-Africa. *Current Opinion in Environmental Science & Health*, *5*, 73–78.

Lafortezza, R., Chen, J., Van Den Bosch, C. K., & Randrup, T. B. (2018). Nature-based solutions for resilient landscapes and cities. *Environmental Research*, *165*, 431–441.

Lee, T. (2013). Global cities and transnational climate change networks. *Global Environmental Politics*, *13*(1), 108–127. https://doi.org/10.1162/GLEP_a_00156.

Lesnikowski, A., Ford, J. D., Biesbroek, R., & Berrang-Ford, L. (2019). A policy mixes approach to conceptualizing and Measuring Climate Change Adaptation Policy. *Climatic Change*, *156*(4), 447–469. https://doi.org/10.1007/s10584-019-02533-3.

Liquete, C., Cid, N., Lanzanova, D., Grizzetti, B., & Reynaud, A. (2016). Perspectives on the link between ecosystem services and biodiversity: The assessment of the nursery function. *Ecological Indicators*, *63*, 249–257.

Lu, P., & Stead, D. (2013). Understanding the notion of resilience in spatial planning: A case study of Rotterdam, the Netherlands. *Cities*, *35*, 200–212. https://doi.org/10.1016/j.cities.2013.06.001.

Maes, J., Liquete, C., Teller, A., Erhard, M., Paracchini, M. L., Barredo, J. I., … Lavalle, C. (2016). An indicator framework for assessing ecosystem services in support of the EU Biodiversity Strategy to 2020. *Ecosystem services*, *17*, 14–23.

Meerow, S., & Newell, J. P. (2019). Urban resilience for whom, what, when, where, and why? *Urban Geography*, *40*, 43–63. https://doi.org/10.4324/9781003130185-3.

Melanidis, M. S., & Hagerman, S. (2022). Competing narratives of nature-based solutions: Leveraging the power of nature or dangerous distraction? *Environmental Science & Policy*, *132*, 273–281.

Ministerie van Financiën. (2022). *Definitief Nederlands Herstel- en Veerkrachtplan*. Ministerie van Financiën.

Murphy, C., Tembo, M., Phiri, A., Yerokun, O., & Grummell, B. (2016). Adapting to climate change in shifting landscapes of belief. *Climatic Change*, *134*, 101–114.

Naef, P. (2022). "100 resilient cities": Addressing urban violence and creating a world of ordinary resilient cities. *Annals of the American Association of Geographers*, *112*(7), 2012–2027. https://doi.org/10.1080/24694452.2022.2038069.

OECD. (n.d.-a). Cities and environment. OECD. https://www.oecd.org/regional/cities/cities-environment.htm.

OECD. (n.d.-b). Resilient cities. OECD. https://www.oecd.org/cfe/resilient-cities.htm.

Otto, A., Kern, K., Haupt, W., Eckersley, P., & Thieken, A. H. (2021). Ranking local climate policy: Assessing the mitigation and adaptation activities of 104 German cities. *Climatic Change, 167,* 1–2. https://doi.org/10.1007/s10584-021-03142-9.

Pan, H., Page, J., Shi, R., Cong, C., Cai, Z., Barthel, S., ... Kalantari, Z. (2023). Contribution of prioritized urban nature-based solutions allocation to carbon neutrality. *Nature Climate Change, 13*(8), 862–870.

Pereira, M. M. (2022). How do public officials learn about policy? A field experiment on policy diffusion. *British Journal of Political Science, 52*(3), 1428–1435. https://doi.org/10.1017/s0007123420000770.

Revi, A., Satterthwaite, D., Aragón-Durand, F., Corfee-Morlot, J., Kiunsi, R. B. R., Pelling, M., ... Solecki, W. (2014). Urban areas. In C. B. Field, V. R. Barros, D. J. Dokken, K. J. Mach, M. D. Mastrandrea, T. E. Bilir, ... L. L. White (Eds.), *Climate change 2014: Impacts, adaptation, and vulnerability: Working Group II contribution to the fifth assessment report of the intergovernmental panel on climate change* (pp. 535–612). Cambrige University Press.

Rockefeller Foundation. (2020, March 25). *100 resilient cities.* Rockefeller Foundation. https://www.rockefellerfoundation.org/100-resilient-cities/.

Seddon, N., et al. (2020). Understanding the value and limits of nature-based solutions to climate change and other global challenges. *Philosophical Transactions of the Royal Society B, 375,* 20190120.

Seddon, N., et al. (2021). Getting the message right on nature-based solutions to climate change. *Global Change Biology, 27,* 1518–1546.

Seddon, N. (2022). Harnessing the potential of nature-based solutions for mitigating and adapting to climate change. *Science, 376*(6600), 1410–1416.

Sharma, M., Sharma, B., Kumar, N., & Kumar, A. (2023). Establishing conceptual components for urban resilience: Taking clues from urbanization through a Planner's lens. *Natural Hazards Review, 24*(1). https://doi.org/10.1061/NHREFO.NHENG-1523.

Staddon, C., Ward, S., De Vito, L., Zuniga-Teran, A., Gerlak, A. K., Schoeman, Y., ... Booth, G. (2018). Contributions of green infrastructure to enhancing urban resilience. *Environment Systems and Decisions, 38,* 330–338.

Steccolini, I., Jones, M. D. S., & Saliterer, I. (Eds.). (2017). *Governmental financial resilience: International perspectives on how local governments face austerity.* Leeds: UK: Emerald Group Publishing.

Tol, R. S. (2018). The economic impacts of climate change. *Review of Environmental Economics and Policy, 12*(1), 4–25. https://doi.org/10.1093/reep/rex027.

Torabi, E., Dedekorkut-Howes, A., & Howes, M. (2021). A framework for using the concept of urban resilience in responding to climate-related disasters. *Cities, 15*(4), 561–583. https://doi.org/10.1080/17535069.2020.1846771.

UN Habitat. (2020). *Climate change vulnerability and risk—A guide for community assessments, action planning and implementation.* UN Habitat. https://unhabitat.org/climate-change-vulnerability-and-risk-a-guide-for-community-assessments-action-planning-and.

UN Habitat. (2021). *City resilience global programme.* UN Habitat. https://unhabitat.org/programme/city-resilience-global-programme.

United Nations. (2016, October 20). *The new urban agenda: Key commitments.* United Nations. https://www.un.org/sustainabledevelopment/blog/2016/10/newurbanagenda/.

United Nations Statistics Division. (2019). *Make cities and human settlements inclusive, safe, resilient and sustainable.* United Nations. https://unstats.un.org/sdgs/report/2019/goal-11/.

Urban Resilience Hub. (2018). *City resilience profiling tool.* Urban Resilience Hub. https://urbanresiliencehub.org/wp-content/uploads/2018/02/CRPT-Guide.pdf.

Wang, C. (2021). Research on sponge city planning based on resilient city concept. *IOP Conference Series: Earth and Environmental Science, 793*(1). https://doi.org/10.1088/1755-1315/793/1/012011.

Wang, L. (2022). Exploring a knowledge map for urban resilience to climate change. *Cities, 131.* https://doi.org/10.1016/j.cities.2022.104048.

Westman, L., & Castán Broto, V. (2022). Urban transformations to keep all the same: The power of ivy discourses. *Antipode, 54,* 1320–1343.

Yi, H., & Liu, I. (2022). Executive leadership, policy tourism, and policy diffusion among local governments. *Public Administration Review, 82*(6), 1024–1041. https://doi.org/10.1111/puar.13529.

Young, A. F., Marengo, J. A., Coelho, J. O. M., Scofield, G. B., de Oliveira Silva, C. C., & Prieto, C. C. (2019). The role of nature-based solutions in disaster risk reduction: The decision maker's perspectives on urban resilience in São Paulo state. *International Journal of Disaster Risk Reduction, 39,* 101219.

Zari, M. P., Kiddle, G. L., Blaschke, P., Gawler, S., & Loubser, D. (2019). Utilising nature-based solutions to increase resilience in Pacific Ocean Cities. *Ecosystem Services, 38,* 100968.

Section 2

Policy approach to promote NbS

Chapter 2.1

Policy, finance, and capacity-building innovations for scaling nature-based solutions

Anna Scolobig[a], Juliette C.G. Martin[b], JoAnne Linnerooth-Bayer[b],
Julia J. Aguilera Rodriguez[a], and Alberto Fresolone[b]
[a]Institute for Environmental Sciences, University of Geneva, Geneva, Switzerland, [b]Equity and Justice Group, International Institute for Applied Systems Analysis, Laxenburg, Austria

1 Introduction

There is growing recognition that nature-based solutions (NbS) offer viable and cost-effective solutions to a broad range of societal challenges (Ozment et al., 2019). NbS can be defined as actions taken "to protect, conserve, restore, sustainably use and manage natural or modified terrestrial, freshwater, coastal and marine ecosystems, which address social, economic and environmental challenges effectively and adaptively, while simultaneously providing human well-being, ecosystem services and resilience and biodiversity benefits" (UNEA, 2022; see White House Council on Environmental Quality and White House Office of Science and Technology Policy, 2022 for an overview of NbS definitions). They encompass a wide range of actions, from planting trees and changing food production methods, to reducing waste and protecting oceans. For instance, restoring native ecosystems can promote healthy soil and vegetation, thereby mitigating risks of floods, erosion, droughts, and landslides by increasing infiltration, water storage, water quality, and the stabilization of slopes and shores (Seddon, 2022). Other examples include the use of NbS over gray solutions for flood protection, trees and parks for cooling residential areas, and forest restoration for reducing wildfire severity (White House Council on Environmental Quality and White House Office of Science and Technology Policy, 2022). NbS implementation can yield significant consequences at multiple levels—for example, local wetlands prevented loss of life and saved communities $625 million in damage during Hurricane Sandy (White House Council on Environmental Quality and White House Office of Science and Technology Policy, 2022).

Nature-Based Solutions in Supporting Sustainable Development Goals
https://doi.org/10.1016/B978-0-443-21782-1.00002-6

129

In 2022, NbS were included for the first time in the Conference of the Parties (COP 27) decision text, which encourages Parties to consider, as appropriate, NbS or ecosystem-based approaches, taking into consideration United Nations Environment Assembly resolution 5/5, for their mitigation and adaptation action while ensuring relevant social and environmental safeguards.

Indeed, NbS have been demonstrated to contribute to the United Nation's 17 Sustainable Development Goals (SDGs) (WWF, 2020), highlighting the underpinning function of nature in achieving sustainable development. For instance, NbS have (among others) been associated with poverty reduction (SDG 1) (Zari et al., 2019), climate action (SDG 13) (Martín et al., 2020), protecting life on land and below water (SDG 14 and 15) (Balzan et al., 2022; O'Leary et al., 2022), and health and well-being (SDG 3). This is amplified by the increasingly understood linkages between all SDGs (Folke et al., 2016), particularly environment-health linkages (Scharlemann et al., 2020), implying great potential for NbS to help simultaneously achieve multiple goals (and related co-benefits).

Furthermore, investing in nature has proven to be a promising strategy that delivers numerous co-benefits with the potential to enhance ecosystem-based disaster risk reduction, increase social and ecological resilience, protect ecosystems, and improve livelihoods, all through the maintenance, restoration, and sustainable use of ecosystems and their services, while addressing other significant policy agendas such as biodiversity and livelihoods (Cohen-Shacham et al., 2016; Palomo et al., 2021; Ruangpan et al., 2020; Seddon, 2022; Seddon et al., 2020).

Thus, while ambition to implement NbS is growing at the international level, their integration into local, regional, and national governance regimes remains problematic. This includes incorporating NbS into regulatory and financial procedures, as well as into risk management, land use, and spatial planning strategies. Additionally, the challenges to NbS implementation must be properly addressed. Prior assessments of NbS feasibility must be carried out, and potential negative consequences and missed opportunities should be considered on a case-by-case basis.

In this chapter, we delve into some of the open issues linked to the realization of the full potential of NbS with the aim to identify the governance, policy, and finance reforms necessary to drive transformative action (IPCC, 2018, p. 558). We build on the results of extensive stakeholder deliberations conducted in the framework of the EC-funded project PHUSICOS (see acknowledgements). In total, 74 NbS experts and knowledgeable stakeholders at the national, European, and international scales were involved in the deliberations of the PHUSICOS Policy Business Forum (PBF) over 4 years. The engagement was conducted using various methods, including interviews, web meetings/e-consultations, and workshops. Specifically, stakeholders deliberated on how to improve the use of existing policies, instruments, and initiatives to better enable the implementation of NbS. They set out to address questions including what changes are needed to catalyze NbS implementation, what role the private and public sectors play, and how they can collaborate, as well as how innovation can be promoted. More precisely, they discussed how to propose new ideas for governance and policy structures that can lead to greater success on NbS acceptance and implementation (for a detailed description of the methodology and results, see Scolobig, Martin, et al., 2023). Three core themes were co-identified with the stakeholders involved and include:

- Governance and policy change/innovations
- Finance innovation and the role of different sectors in mainstreaming NbS
- Private sector capacity building

These themes represent the thread of this chapter. In the following sections, after a description of the methodology to engage with stakeholders, we look at the state of the art and key results for each core theme, including recommendations and innovations suggested during the deliberations. We conclude with future perspectives for NbS transformative action.

2 Stakeholder deliberations

The methodology for the stakeholder deliberations was grounded on a triangulation of different social science methods (Bryman, 2012; Silverman, 2010), namely semistructured interviews, surveys, and workshops.

Semistructured interviews and questionnaire surveys were carried out as preparatory work to identify relevant themes for professionals working on NbS governance, collect information on key challenges related to those themes, and better understand perspectives on opportunities and gaps. The semistructured interviews were started several months before the workshops, with a total of 15 interviews conducted.

A stakeholder database was established and used to contact the workshop participants, together with snowballing. Three workshops have been organized for the PHUSICOS PBF: each attended by 18–36 participants.

The workshop participants had a variety of backgrounds, such as engineering, natural sciences, social sciences, and business. Most participants work in policy and/or DRR, and their interest in NbS had been motivated primarily by the urgency of the climate crisis and the co-benefits of NbS, which aim to address many societal challenges.

The workshops focused on the core themes co-identified with the stakeholders (see Section 1).

For each workshop, we identified core questions based on a literature review of the selected theme, the preparatory interview results, and short questionnaires sent to all participants. While the first theme was chosen by the research team, subsequent themes emerged from gaps identified during the workshop discussions. Table 1 provides an overview of the themes and questions addressed in the PBFs.

TABLE 1 Key questions addressed in the PBFs.

PBF workshop theme	Key questions
Governance innovation for NbS	How can NbS be mainstreamed into European DRR policy agendas?
	Will the EU Green Deal result in changing NbS policy?
	Are new directives and frameworks needed at the European level?
	If so, which ones (e.g., can the Common Agricultural Policy include incentives for farmers to adopt NbS-targeted DRR)?
	Are there cross-country differences in the NbS policies and instruments (for DRR) in Europe?
	What are the main barriers to implement new national policies?
	Are some countries more advanced than others? Why so?
	Can NbS effectiveness be measured? How?
	Do we need new NbS national regulation? If so, on what (e.g., is landslide mitigation in need of legislation about effects of root systems, including quantification)?
The role of public and private sectors in mainstreaming NbS	How do private sector organizations define priorities for NbS action?
	What barriers do they encounter?
	Which of the many innovative market instruments holds promise for NbS scaling?
	How can public banks provide more funding for NbS?
	How can public authorities foster synergies, not only between policy sectors but also between the public and private sectors?
	Are there policies in place that hinder NbS financing?
	How can we significantly increase funding for the public sector NbS agenda (e.g., COVID-19 recovery fund)?
NbS private sector upscaling	What nature-based enterprise capacities are lacking (e.g., on the part of NbS design and construction) that could enable the upscaling of NbS? How can these capacities be built, and by whom? What are the priorities?

TABLE 1 Key questions addressed in the PBFs—cont'd

PBF workshop theme	Key questions
	What are the main existing policies, mechanisms, and/or resources that could help overcome capacity gaps to support NbS upscaling?
	What are the main risks associated with the design, construction, operation, and maintenance of NbS? Who or what institutions can be held responsible for such risks?
	In the event of failure of NbS to protect material assets and human lives (e.g., in DRR interventions), who can be held responsible or even liable? Have insurers played a role (or can they play a role) in de-risking NbS performance?
	What policies, guidelines, and regulations have been (or should be) put in place to alleviate NbS failure concerns?

After the workshops, a synthesis was prepared based on the recording of the preparatory semistructured interviews, the results of the surveys administered to the participants before the workshops, and the workshop recordings and notes. To analyze the data, we used standard qualitative data analysis techniques based on thematic analysis (Bryman, 2012; Silverman, 2010). More precisely, we identified the core themes emerging from the deliberations, accompanied by selected significant quotes. To ensure that our syntheses reflect participant views, we shared the draft with the participants and gave them the opportunity to comment, revise, or add further contributions. The final versions of the three syntheses (Aguilera Rodriguez et al., 2022; Scolobig et al., 2020, 2021) serve as background for this chapter.

The forum agendas, keynote/presentation abstracts, and core questions of the thematic sessions can be found in Scolobig, Aguilera, et al. (2023). In the following sections, we present the state of the art for each core theme and summarize the key issues that emerged during the stakeholder deliberations, followed by a list of recommendations and suggested innovations.

3   Policy and governance

Analyses of NbS policies have examined if and how regional, national, and international policy frameworks address the concept of NbS (Davies et al., 2021). For example, at the European level, these analyses revealed that the

European Commission (EC) is investing considerably in NbS and green growth, with the goal of positioning Europe as a leader of "innovation with nature" (Davies et al., 2021; EEA, 2021). Partially because of this push, NbS are seen in several different policy domains, including adaptation, disaster risk management, research and innovation, biodiversity, health, and water retention. For example, NbS are included in the new EU Strategy on Adaptation to Climate Change, in which they are considered essential for increasing climate resilience and sustaining healthy water, oceans, and soils (EC, 2021). The recently released EU Biodiversity Strategy and the EU Forest Strategy, both for 2030 (EC, 2020), are key pillars of the ambitious European Green Deal (EC, 2019). Both strategies also rely on NbS to increase climate resilience and preserve and restore ecosystem integrity.

NbS are also increasingly featured in other EU policies and strategies beyond conservation and environmental protection. For example, the Farm to Fork Strategy (EC, 2020) promotes the use of NbS, such as agroecology practices, to transition to a sustainable food system. Likewise, an analysis of EU-funded projects emphasized the direct relevance of the EU Flood Directive and Water Framework Directive for NbS implementation (Vojinovic, 2020). Meanwhile, the EU Floods Directive has been successful in inspiring at least 26 Member States to include natural water retention measures, a form of NbS, in their plans.

Despite these advancements, broader adoption is needed to meet the ambitious United Nations goal of tripling investments in NbS (Linnerooth-Bayer, Martin, Fresolone-Caparrós, Scolobig, & Aguilera Rodriguez, 2023). Indeed, current analyses of NbS policies reveal a predominantly nonbinding instrument mix for NbS and related concepts. Policies are largely based on voluntary action and often lack quantitative and measurable targets for NbS deployment and quality evaluation (Davis et al., 2018). Many policies explicitly describe the benefits of nature, allude to their potential to address societal challenges, and encourage action to adopt or promote such measures; however, they often do not go further than encouragement, failing to set standards or mandate supportive action, policy, or financial instruments (Davies et al., 2021). Furthermore, the complex mosaic of policy instruments addressing NbS can lead to confusion among decision-makers, fragmented governance, and, eventually, policy stalemates (EEA, 2021; Trémolet et al., 2019). For example, there is a lack of coherence both in EU and sectoral policies relating to NbS (EEA, 2021; Trémolet et al., 2019). Further alignment of sectoral policy instruments is thus needed to facilitate cross-sectoral (and by extension polycentric) governance arrangements for NbS (EEA, 2021).

At the national level, multiple policy instruments explicitly acknowledge NbS and related concepts, occasionally including them in their strategic objectives (e.g., German Green and White Papers on Urban Green, Federal Strategy on green infrastructures, Spanish National Natural Heritage and Biodiversity Law), but there are still glaring gaps in the identification of quantitative and measurable targets (EEA, 2021) and NbS are not mainstreamed into existing national policies. Indeed, while NbS are featured in a large proportion of nationally determined contributions and national adaptation plans (Seddon et al., 2020), this has not yet translated into their widespread implementation.

This implementation gap—the discrepancy between NbS ambitions and on-the-ground implementation—has been noted in several studies and maps well onto the "policy-action" gap identified in the latest IPCC report (Dodman et al., 2022). The policy-action gap arises when administrative, communication, financial, and other organizational blockages and inertia interrupt policy implementation, the intent of political leadership, and the delivery of adaptation interventions on the ground (Dodman et al., 2022: 969). For example, Runhaar et al. (2018) find that in many European countries, the lagging of policy integration of adaptation strategies, including NbS (see UNEP, 2022a), is likely due to a lack of both long-term political commitment and effective coordination between key stakeholders. In Switzerland, the adaptation gap was also attributed to a lack of political commitment at the national and cantonal levels (Braunschweiger & Pütz, 2021). Frantzeskaki et al. (2020) highlight the need for national policies to foster institutional spaces for collaborative learning and targeted capacity-building programs (see also Section 5).

The need for NbS national policies is parallel with the need for maturation of NbS governance, which goes beyond "government" and the arrangements it encompasses to include a network of state, nonstate, and business actors in the process of deciding on and implementing NbS policies (Lemos & Agrawal, 2006; Lupp & Zingraff-hamed, 2021; Martin et al., 2019, 2021; Steurer, 2013). As such, governance also encompasses the social, legal, institutional, political, and financial conditions through which NbS are implemented (Bernardi et al., 2019). Governance innovation may therefore foster the prioritization of NbS as either a complement or alternative to traditional gray infrastructure approaches, promote a wider uptake, and increase public awareness due to collaborative design and implementation processes (Seddon et al., 2020).

The diverse challenges to NbS governance depend on a complex array of conditions, including policy frameworks, political regimes, and institutions, as well as local socioeconomic contexts (Faivre et al., 2017; Naumann & Davis, 2020). Nevertheless, several challenges are shared, including the need to exit gray-solution path dependency and balance the demands of nature preservation/restoration with urban and rural growth and existing institutional settings. For example, local governments face persistent planning challenges regarding the integration of new knowledge and new governance approaches due to continued silo thinking, resistance to novelty, policy fragmentation, and lack of budget (Mahmoud et al., 2022; Martin et al., 2021). The short-term nature of public and private sector decision-making and budget allocation hinders longer-term planning, management, and maintenance of NbS, which are essential to address climate change and biodiversity goals in a sustainable manner (Seddon et al., 2020). Most importantly, NbS may take a long time to show benefits (Solheim et al., 2021). A systematic review of barriers to NbS implementation is provided in Martin et al. (forthcoming).

In summary, gaps concerning the identification of key innovations that decision-makers and practitioners can adapt to NbS implementation from a governance perspective persist (Frantzeskaki, 2019; Frantzeskaki et al., 2020). This includes identifying relevant policy mechanisms along with levers for institutional reform that can better enable NbS implementation and upscaling (Fedele et al., 2019).

3.1 Stakeholder deliberations

In this section, we summarize the key themes linked to NbS governance and policy innovation that emerged during the stakeholder deliberations described in Section 2.

Mandatory policy instruments: To date, many EU policy mechanisms enabling NbS remain voluntary and thus have no legal obligations to comply. Mandatory policy instruments, e.g., making NbS compulsory elements of landscape planning, could be further promoted. Examples include the protection of a proportion of land for forest cover in EU Member States or the inclusion of legally binding nature restoration targets, as in the proposed EU Nature Restoration Law.

Standards: Clear formal standards are increasingly recognized as key elements guiding the design and implementation of NbS. By outlining a set of evidence-based criteria, standards assist in ensuring the quality, safety, and efficiency of interventions while supporting their long-term sustainability and minimizing the possibility of unwanted social and environmental impacts. In some cases, given the contextual sensitivity of NbS, tailored standards for specific types of solutions may be called for.

Co-benefit evaluation: Implementing new NbS policies within Member States is essential to promoting the NbS agenda. Yet, further research is needed to support the quantification of NbS co-benefits, prove their transferability, and verify their effectiveness at larger scales. This information can then be used within innovative methods to improve the visibility of NbS investment co-benefits for regulatory and funding decisions. In the meantime, qualitative targets can be used until tools for co-benefit quantification are available. While this does not solve the numerous challenges related to the difficulty of quantifying nonmarket benefits and interconnected impacts, qualitative targets would provide decision-makers with much-needed evidence on the viability of NbS.

Policy synergies and trade-offs: There is potential to link NbS policies to other policies related to well-being and preventative health care, biodiversity, green infrastructure, transport, and mobility. Cross-sectoral integration of NbS and related concepts is key to generate concrete implementation action. One integrated NbS policy example could be the co-development of joint biodiversity and climate plans at regional or national level. However, trade-offs are likely to occur and should not be overlooked. For example, when seeking both biodiversity conservation and human well-being goals simultaneously, such as promoting NbS while providing opportunities for new developments such as housing.

Integrated and inclusive bottom-up NbS planning: Few guidelines exist to support NbS implementation at small scales, often leading to the growth of single and isolated solutions that are not strategically integrated on a larger scale. This is problematic because NbS often require extensive and expensive land acquisition on a grand scale. To address this challenge, integrated and inclusive bottom-up NbS planning should be promoted systematically. Innovations to empower the local level include NbS knowledge hubs, close collaboration with NbS project ambassadors/local champions, and promotion of cross-sector and cross-scale institutional mechanisms.

Divestments from gray infrastructures and nature-negative projects: The unbalanced distribution of funding for green and gray measures, as well as greenwashing, are key issues hindering NbS implementation (Browder et al., 2019). Developing policy instruments that allow to divest from gray infrastructures would help to overcome this imbalance. One reported example is encouraging or even requiring financial institutions to divest from nature-negative activities and to invest in nature-positive activities. This would redirect substantial amounts of financing to NbS. However, these should be combined with transparent ways to identify nature-positive solutions without greenwashing. Another solution is switching the burden of proof to gray measures, e.g., making consideration of NbS a requirement for infrastructure projects (see also Linnerooth-Bayer, Martin, Fresolone-Caparrós, Scolobig, Aguilera Rodriguez, Solheim, et al., 2023). Despite the need to divest from gray infrastructures, in some cases (especially when life losses are at stake), hybrid or gray solutions may still be the preferred option to guarantee high-security standards.

Green and gray funding streams and land acquisition: Although the capital costs of NbS can be less than for gray infrastructure, there are still significant costs in their implementation, including expensive land acquisition, which make upscaling difficult. This barrier could be addressed with innovative funding mechanisms (e.g., Payment for Ecosystem Services). The lack of balance between funding for green and gray solutions, due in large part to separate funding streams, further complicates matters. In this case, a potential solution is to merge complementary funding streams into single programs (e.g., disaster risk reduction plans or water management plans) that prioritize NbS approaches.

3.2 Recommendations and suggested innovations

To address the challenges outlined in the previous section, one key recommendation is to update policies and promote mandatory policy instruments, for example, by making NbS compulsory elements of landscape planning, listing the evaluation of NbS options as a requirement of infrastructure projects, or streamlining simplified NbS approval procedures. This last could be done through, for example, the introduction of self-certification schemes. Other suggested innovations include:

- Enforcing legally binding targets, e.g., reservation of a proportion of land for forest cover in Member States as in the proposed EU Nature Restoration Law.
- Switching the burden of proof to gray measures, e.g., make NbS compulsory as elements of landscape planning and/or require stringent documentation of the long-term negative impacts of gray infrastructure.
- Fostering policy synergies by linking NbS policies to well-being and preventative healthcare policies or to green infrastructure, transport, and mobility policies, e.g., joint biodiversity, and climate plans at regional or national level.
- Promoting cross-sectoral and multilevel collaboration and polycentric governance arrangements, e.g., cross-sectoral secretariats for NbS strategies.
- Divesting from nature-negative projects and investing in nature-positive activities, e.g., expansion of EU taxonomy for sustainable finance.

4   Finance and the role of the private sector

NbS financing is an area of growing interest to explore. While it is estimated that an average of USD 154 billion was invested worldwide annually in NbS as of 2022 (UNEP, 2022b), this still represents a small percentage of the USD 579 billion spent annually on climate finance in general (UNEP, 2021). This gap is reported as a key barrier for reaching, for example, climate mitigation and adaptation goals within urban landscapes (Toxopeus & Polzin, 2021), yet it nonetheless opens a window of opportunity for scaling up, particularly for the private sector. NbS financing refers to securing funds for NbS planning, implementation or/and maintenance, and operation (McQuaid, 2020). Currently, NbS financing relies heavily on public financing (83% of total NbS investment), with just 17% coming from the private sector (UNEP, 2022a). Increasing private sector investment would provide a pivotal opportunity to boost NbS deployment.

Two main drivers of private NbS financing have been identified by the World Business Council for Sustainable Development: voluntary actions and regulatory compliance. Voluntary investments are heavily influenced by customers, investors, and employees to act on climate change, while regulatory initiatives can potentially generate funds, for example, through the adoption of carbon taxes (World Business Council for Sustainable Development, 2019). However, increasing private investment faces a number of obstacles due to the complexity of NbS solutions, the specificity of the contexts in which they are applied, and the multiple benefits and stakeholders involved (Mayor et al., 2021). Barriers include difficulties in capturing revenues given the nature of most NbS benefits as public goods, challenges in valuing and accounting for multiple benefits and co-benefits, a lack of predictable long-term revenue streams, and the frequent need for long-term financing (Hagedoorn et al., 2021; Toxopeus & Polzin, 2021). Financial risks to the private sector can be reduced through third-party guarantees, new insurance products, blended financing, and other strategies (EIB, 2020).

Diverse economic and financial mechanisms exist that can encourage further private uptake of NbS, including carbon markets, green financial products (e.g., green bonds), payments for ecosystem services, reduced insurance premiums, and public and private funds (e.g., the Adaptation Fund, the European Agricultural Fund for Rural Development, the European Regional Development Fund, Climate Asset Management, and Mirova Natural Capital) (UNEP, 2022b). In the case of NbS for climate change adaptation and disaster risk reduction, support can take the form of incentives (e.g., subsidies and payments), disincentives (e.g., taxes or charges), or risk-financing schemes (e.g., insurance and other risk transfer mechanisms) (Calliari et al., 2022).

At the same time, there is unprecedented political momentum and windows of opportunities for scaling up NbS with increasing financing opportunities and more solid foundations for implementation provided by research and practice (Davies et al., 2021). For example, the EU Biodiversity Strategy outlines an ambition to unlock at least €20 billion a year for spending on nature, and the Natural Capital Financing Facility of the European Investment Bank allocated €250 billion to the EU Green Deal. Moreover, NbS are playing a role in making the post-COVID-19 recovery and the implementation of the EU Recovery and Resilience Facility green, healthy, just, and equitable (European Commission, n.d.). Innovative investment models are also emerging, such as those of venture funds specifically focused on biodiversity and banks with funds targeted for natural capital (Surminski et al., 2022).

Other ways to reduce financial deficits and requirements include voluntary time contributions from citizens (in the form of unpaid work) to cover part of the labor needs for NbS implementation (Hagedoorn et al., 2021). This strategy is considered to have substantial potential for community-led interventions and for developing countries, where the availability of public funding is generally lower compared to developed nations and where NbS are often donor-driven (Bhattarai et al., 2021).

Regardless of the financing source, there is a consensus in the literature that capital directed to NbS is currently insufficient and that sources other than the public sector must be engaged (UNEP, 2021, 2022a, 2022b). Alongside the economic incentives described above, other supportive strategies are essential to facilitate an enabling environment (EIB, 2020; Mayor et al., 2021; Toxopeus & Polzin, 2021; UNEP, 2021). Technical assistance to businesses to identify bankable products and increasing business skills and knowledge on NbS (EIB, 2020), strong public and political leadership, support for the co-design of solutions (Schröter et al., 2022), the creation of standard metrics, baselines, and common characteristics for NbS, and the establishment and integration of environmental and social safeguards in implementation (UNEP, 2021) are among the most prominent.

4.1   Stakeholder deliberations

A number of key points on financing models emerged during the stakeholder deliberations, including co-financing options, barriers to private financing, de-risking, and insuring NbS.

Co-financing options: For many NbS investments, private investors have difficulty capturing revenues that generate a sufficient return on their investment, which limits interest from traditional financial institutions. To address this, co-financing options can provide incentives to private investors, including subsidies and tax rebates for NbS investments. Alternatively, participants pointed out that NbS projects can be co-developed and co-financed as public-private partnerships (PPPs), possibly including the public sector, financial institutions, and private enterprises. Other promising trend is the upscaling of innovative private and blended financing models such as collective investment schemes and stewardship schemes.

Barriers to private financing: Private firms are more likely to opt for NbS investment if the direct benefits, or revenues, to the enterprise outweigh the investment costs more than other options. Private individuals or entrepreneurs might then fully finance NbS with their own capital or with credit from a private or public bank such as the European Bank for Reconstruction and Development (EBRD) or the European Investment Bank (EIB). However, as noted by participants, identifying "bankable" projects presents a formidable challenge. Other barriers to private financing include a lack of information on NbS effectiveness, unfavorable regulations, a lack of awareness on behalf of enterprises, path dependency, and difficulties in shifting norms and culture from traditional gray solutions. Other obstacles for private investment in NbS are lack of knowledge and uncertainty in the effectiveness of such types of measures, yet these might be overcome by deploying innovative financial instruments to de-risk projects (e.g., private or public insurance and provision of public guarantees).

Innovative financial instruments: To de-risk NbS projects, innovative financial instruments can be arranged. In addition to insurance and government guarantees, another innovation is the so-called "resilience bond," a variation on green bonds that seeks to raise capital specifically for climate-resilient investments. Resilience bonds refer to a form of catastrophe bonds that link insurance premiums to resilience projects in order to monetize avoided losses through a rebate structure.

Insuring NbS: Insurance and reinsurance schemes applied to NbS could play a significant role in spreading the risks associated with their design, construction, and performance, covering, for example, risks of delays and budget overruns, as well as liability over their performance. Insurance can reduce the risks of implementing NbS, notably in high-risk scenarios, such as NbS for

DRR and climate change adaptation, which traditionally employ gray solutions. Other risk-reduction alternatives include the participation of governments as insurers or reinsurers to absorb a portion of the risk, as well as the possibility of introducing community-based insurance schemes. However, to do so, clearer design standards and the accumulation of operational experience and data to support the effective development of risk assessments are still required.

4.2   Recommendations and suggested innovations

One key recommendation is to unlock public and private funding for NbS design and implementation. In this respect, the following innovations may be considered:

- Merging complementary funding streams (green and gray) into single programs that prioritize NbS.
- Developing innovative financing mechanisms, e.g., resilience bonds or payments for ecosystem services to address land availability problems.
- Promoting co-financing options, e.g., subsidies and tax rebates for NbS investments.
- Promoting partnerships between the public sector, financial institutions, and private enterprises and applying blended financing models, e.g., collective investment and stewardship schemes.
- De-risking NbS, e.g., ensuring NbS to transfer risk of project design, construction, and loss-and-damage from extreme weather; providing government guarantees for operational and liability risk; and innovating with community-based insurance schemes.

5   Capacity building

The NbS implementation process involves a wide range of stakeholders, making it crucial for each of them to have the necessary skills and knowledge to play their role and ensure the optimal performance of solutions (Mabon et al., 2022). Public authorities are typically the primary actors behind most phases of NbS implementation, followed by civil society and private service providers, or contractors (Zingraff-Hamed et al., 2020).

NbS contractors are usually hired for specific tasks over a set period and may be involved in various phases of the solutions, including planning, design, construction, or maintenance. They encompass a diversity of firms such as landscape architects, ecologists, consultants, engineers, or construction company employees (Kooijman et al., 2021) who work at what can be classified as nature-based enterprises, engaged in an economic activity that directly or indirectly "uses the sustainability of nature as a core element of their product/ service offering" (McQuaid et al., 2020). Contractors contribute their expertise and resources to less-experienced project authorities for NbS implementation (Tilt & Ries, 2021).

However, recent studies demonstrate that the lack of experience and knowledge of private NbS service providers is a major barrier currently affecting NbS implementation (Kuhlicke & Plavsic, 2021; Martin et al., forthcoming; Solheim et al., 2021). NbS knowledge gaps vary given that these solutions require a multidisciplinary skill set and competencies spanning from academic, to technical and commercial (ILO et al., 2022). For example, many nature-based enterprises with strong technical/ecological skills may have poor business and communication capabilities (McQuaid et al., 2021). These difficulties are particularly problematic in cases where companies with little or no prior experience in NbS are recruited, as frequently occurs with construction companies exclusively dedicated to the deployment of gray infrastructure and which, for various reasons (e.g., impossibility to find other skilled local providers) are given the task of building NbS (Linnerooth-Bayer, Martin, Fresolone-Caparrós, Scolobig, Aguilera Rodriguez, Solheim, et al., 2023; Mačiulytė & Durieux, 2020).

The consequences of insufficient supplier experience in NbS implementation include a low number of bids for projects (McQuaid et al., 2021), poor data collection, difficulties in conducting robust project evaluations to ensure effectiveness, as well as negative impacts on the cost and quality of measures deployed (Mačiulytė & Durieux, 2020). These issues not only affect contractor performance but also overall NbS project results and the perception decision-makers have of such projects. Moreover, outcomes can also be impacted by a still-immature NbS policy regime that reduces the capacity to enforce the implementation of such measures.

Developing appropriate skills could assist workers and enterprises in implementing NbS, thus simultaneously creating new and fairly compensated employment opportunities (ILO et al., 2022). However, just and equitable learning opportunities for those in charge of implementing solutions on the ground are required (Mabon et al., 2022). Clear standards and safeguards to guide the design and implementation of NbS are reportedly much needed (UNEP, 2021), as well as exchange platforms to facilitate stakeholder communication and organization. Such platforms are said to be important enablers of NbS upscaling (Fastenrath et al., 2020). These observations are based primarily on the study of past projects and the general exploration of facilitating and hindering factors for NbS. Further research into contractors' perceptions of the limitations they experience and the factors they judge necessary to overcome them is necessary. Linnerooth-Bayer, Martin, Fresolone-Caparrós, Scolobig, Aguilera Rodriguez, Solheim, et al. (2023) provide an analysis of the experiences and needs of NbS professionals working in the provision of NbS services, including designers, construction companies, and consulting firms.

5.1   Stakeholder deliberations

In this section, we summarize the key themes linked to NbS capacity building that emerged during the stakeholder deliberations.

Creation of facilities for the private sector: Some of the major challenges for NbS implementation are related to the lack of specialized companies and the low supply of expertise for their planning, construction, and maintenance. Many companies lack practical experience in NbS implementation as well as basic business and marketing skills. To support private sector development, the creation of an NbS Project Preparation Facility (PPF) at the local and/or national scale and a user-friendly benefit and co-benefit catalog for the private sector are potential options. PPFs can assist nature-based enterprises from the conception, design, or scoping phases through to project execution and closure by providing support in the formulation of feasibility studies, risk assessments, tender applications, etc. Additionally, accelerator programs could offer start-ups growth and learning opportunities through intensive funding and mentoring.

Capacity-building for NbS contractors: The development of easily accessible learning tools is one potential strategy for improving the skills and capabilities of NbS contractors, including consultants, designers, constructors, and maintenance companies. Backed by experience, training courses and seminars, including online versions, can benefit contractors by providing evidence-based information on the potential uses, limitations, and maintenance of NbS in a wide range of scenarios. Moreover, such tools can also serve to disseminate existing or emerging guidelines, promoting best practices more effectively and providing clear recommendations. Improving contractor competencies may significantly reduce the probability of NbS failure, positively influencing public perceptions and increasing the likelihood of replication. Finally, training and awareness-raising on already available guidelines and standards (e.g., the IUCN Global Standard for NbS) are paramount, along with the constant improvement of such tools.

5.2 Recommendations and suggested innovations

To support private sector upscaling and build capacities, the following options were suggested during the stakeholder consultations:

- Creating an NbS Project Preparation Facility (PPF) at the local and/or national level by providing support to nature-based enterprises, e.g., in the formulation of feasibility studies, risk assessments, and tender applications.
- Creating user-friendly benefit and co-benefit catalogs for the private sector.
- Developing new educational and training programs specific to NbS design and implementation, including available guidelines and standards, e.g., the IUCN Global Standard for NbS.
- Using innovative learning tools such as virtual-reality learning platforms.
- Incorporate multidisciplinary competencies into NbS curricula, including courses on NbS legislation.
- Creating accelerator programs/mentoring initiatives to provide start-ups with growth and learning opportunities through intensive funding and mentoring for a brief period.
- Developing NbS Hubs to facilitate communities of practice for NbS contractors, involving the public, academia, and civil society.

6   Conclusion

For NbS implementation to become more widespread and effective, it is essential to transform governance structures and create appropriate legal, institutional, political, and financial conditions. Although ambition at the international level is growing, policy development at national and regional levels, as well as NbS upscaling at the local levels are often problematic. Current policies, regulations, and path dependency from gray solutions frequently create substantial hurdles (Martin et al., 2023). Often, agencies and communities find funds for NbS insufficient or financing processes difficult to navigate or access. Additionally, the distribution of funding for green and gray measures is still very unequal and the private sector expertise to design and build NbS lags in many regions (Linnerooth-Bayer, Martin, Fresolone-Caparrós, Scolobig, & Aguilera Rodriguez, 2023). Furthermore, technical challenges, information deficits, and uncertainties related to NbS effectiveness can also pose formidable obstacles, together with a lack of quantitative measurable targets for NbS deployment and quality (White House Council on Environmental Quality and White House Office of Science and Technology Policy, 2022).

In this chapter, we present the key results of stakeholder consultations, offering new ideas for governance and policy structures that can enhance NbS acceptance and implementation. Over 70 stakeholders were involved in deliberations through interviews, web meetings/e-consultations, and workshops. Fig. 1 provides an overview of the main results of the deliberations, with a focus on innovations for scaling nature-based solutions.

We summarize here some key messages and highlight knowledge gaps and new research directions.

The first key message is the promotion of mandatory policy instruments. This includes the enforcement of legally binding targets and the simplification of NbS approval procedures. Further, measures could include fostering policy synergies.

The second message is unlocking public and private funding to enable NbS investments, merging complementary funding streams into single programs that prioritize NbS, and promoting innovative financing mechanisms, such as payment for ecosystem services. Meanwhile, divesting from nature-negative projects is as important as enabling NbS investments.

The third one is capacity building through the creation of NbS project preparation facilities, accelerator programs/mentoring initiatives, user-friendly benefit/co-benefit catalogs for the private sector, and the creation of communities of practice for NbS contractors with the public, academia, and civil society.

Likewise, strengthening the knowledge base is also essential to build capacities. New research directions and innovative methodologies include the improvement of NbS monitoring, assessment, and co-benefit evaluation. Thus, future research should strengthen the evidence on NbS effectiveness, contribute

POLICY

- Enforce legally binding targets

- Switch the burden of proof to grey measures

- Foster policy synergies by linking NBS policies to other sectoral policies (e.g. on wellbeing)

- Promote cross-sectoral and multi-level collaboration and polycentric governance arrangements

FINANCING

- Merge complementary funding streams (green and grey)

- Develop innovative financing mechanisms

- Promote co-financing options

- Promote partnerships between the public sector, financial institutions, and private enterprises

- De-risking NBS through insurance

- Divest from nature-negative projects and invest in nature-positive activities

CAPACITIES

- Create an NBS Project Preparation Facility (PPF) supporting nature-based enterprises

- Create user-friendly benefit and co-benefit catalogues for the private sector

- Develop new educational and training programmes for NBS design and implementation

- Use innovative learning tools

- Integrate multidisciplinary competencies in NBS curricula

- Create accelerator programmes/mentoring

- Create communities of practice for NBS contractors through NBS Hubs

FIG. 1 Innovations for scaling nature-based solutions.

to the development of formal standards, including building codes and insurance regulations based on the evidence created, improving co-benefit evaluation and quantification, and developing tools to compare NbS, hybrid, and conventional solutions. Moreover, it would be beneficial to provide guidelines and roadmaps for NbS implementation tailored to cities/regions at various stages of policy maturity or facing distinct financial/social environments.

We hope that these recommendations and suggested innovations will contribute to fostering debate and supporting the uptake of NbS as key options in fighting climate change, biodiversity loss, and reducing disaster risk.

Acknowledgments

The work described in this chapter was supported by the European Community through the grant to the PHUSICOS Project (https://phusicos.eu/). PHUSICOS was an Innovation Action project funded by the EU Horizon 2020 research and innovation program (Grant agreement No. 776681; 2018–23). The main objective of PHUSICOS was to demonstrate that NbS for reducing natural hazard risk of extreme weather events in rural mountain landscapes are technically viable, cost-effective, and implementable at the regional scale. The chapter reflects the authors' views and not those of the European Community. Neither the European Community nor any member of the PHUSICOS Consortium is liable for any use of the information in this chapter. We wish to thank all the colleagues, including all PHUSICOS coordinator, partners, and persons who provided us with professional advice and collaboration for setting up this research and the Policy Business Forum (PBF). And we would like to express our gratitude to the 74 participants and presenters who dedicated their time to the interviews, surveys, workshops, and all the activities linked to the PBF.

References

Aguilera Rodriguez, J. J., Scolobig, A., Linnerooth-Bayer, J., Martin, J. G. C., Fresolone-Caparrós, A., Irshaid, J., ... Pakarinen, N. (2022). *NBS private sector upscaling and capacity building. Third policy business forum synthesis. PHUSICOS project, according to nature. Nature based solutions to reduce risks in mountain landscapes, EC H2020 Programme.* Retrieved from https://phusicos.eu/.

Balzan, M. V., Geneletti, D., Grace, M., De Santis, L., Tomaskinova, J., Reddington, H., ... Collier, M. (2022). Assessing nature-based solutions uptake in a Mediterranean climate: Insights from the case-study of Malta. *Nature-Based Solutions, 2*, 100029.

Bernardi, A., Enzi, S., Mesimäki, M., Lehvävirta, S., Jurik, J., Kolokotsa, D., Gobakis, K., van Rompaey, S., et al. (2019). *Barriers landscape and decision-making hierarchy for the sustainable urbanisation in cities via NBS.* Bru: ThinkNature Project.

Bhattarai, S., Regmi, B. R., Pant, B., Uprety, D. R., & Maraseni, T. (2021). Sustaining ecosystem-based adaptation: The lessons from policy and practices in Nepal. *Land Use Policy, 104*, 105391. https://doi.org/10.1016/j.landusepol.2021.105391.

Braunschweiger, D., & Pütz, M. (2021). Climate adaptation in practice: How mainstreaming strategies matter for policy integration. *Environmental Policy and Governance, 31*(4), 361–373. https://doi.org/10.1002/EET.1936.

Browder, G., Ozment, S., Bescos, I. R., & Gartner, T. (2019). *Integrating green and gray: Creating next generation infrastructure.* Washington: World Resources Institute and World Bank Group.

Bryman, A. (2012). *Social research methods* (4th ed.). New York: Oxford University Press, ISBN:978-0-19-958805-3.

Calliari, E., Castellari, S., Davis, M., Linnerooth Bayer, J., Martin, J., Mysiak, J., ... Zandersen, M. (2022). Building climate resilience through nature-based solutions: A review of enabling knowledge, finance and governance frameworks. *Climate Risk Management, 37*. Retrieved from https://www.sciencedirect.com/science/article/pii/S2212096322000572.

Cohen-Shacham, E., Walters, G., Janzen, C., & Maginnis, S. (2016). *Nature-based solutions to address global societal challenges*. Gland, Switzerland: IUCN.

Davies, C., Chen, W., Sanesi, G., & Lafortezza, R. (2021). The European Union roadmap for implementing nature-based solutions: A review. *Environmental Science and Policy, 121* (March), 49–67. https://doi.org/10.1016/j.envsci.2021.03.018.

Davis, M., Abhold, K., Mederake, L., & Knoblauch, D. (2018). *Nature-based solutions in European and National Policy Frameworks*. NATURVATION Project, 52.

Dodman, D., Hayward, B., Pelling, M., Castan Broto, V., Chow, W., Chu, E., ... Ziervogel, G. (2022). Cities, settlements and key infrastructure. In H.-O. Pörtner, D. C. Roberts, M. Tignor, E. S. Poloczanska, K. Mintenbeck, A. Alegría, ... B. Rama (Eds.), *Climate change 2022: Impacts, adaptation and vulnerability. Contribution of working group II to the sixth assessment report of the intergovernmental panel on climate change* (pp. 907–1040). Cambridge, UK and New York, NY, USA: Cambridge University Press. https://doi.org/10.1017/9781009325844.008.

EIB, European Investment Bank. (2020). *Investing in nature: Financing conservation and nature-based solutions*. European Investment Bank. Retrieved from https://www.eib.org/attachments/pj/ncff-invest-nature-report-en.pdf.

European Commission. (2019). *Communication from the commission to the European Parliament, the Council, the European Economic and Social Committee and the Committee of the Regions— The European Green Deal*. Brussels: European Commission. https://doi.org/10.1017/CBO9781107415324.004.

European Commission. (2020). *EU biodiversity strategy for 2030 (COM(2020) 380 final)*. Brussels: European Commission. Retrieved from https://eur-lex.europa.eu/resource.html?uri=cellar:a3c806a6-9ab3-11ea-9d2d-01aa75ed71a1.0001.02/DOC_1&format=PDF.

European Commission. (2021). *Forging a climate-resilient Europe: The new EU Strategy on Adaptation to Climate Change*. Retrieved from https://eur-lex.europa.eu/legal-content/EN/TXT/?uri=COM:2021:82:FIN.

European Commission. (n.d.). Recovery and resilience facility. The recovery and resilience facility. Retrieved from: https://commission.europa.eu/business-economy-euro/economic-recovery/recovery-and-resilience-facility_en.

European Environment Agency. (2021). *Nature-based solutions in Europe: Policy, knowledge and practice for climate change adaptation and disaster risk reduction*. EEA. https://doi.org/10.2800/919315.

Faivre, N., Fritz, M., Freitas, T., de Boissezon, B., & Vandewoestijne, S. (2017). Nature-based solutions in the EU: Innovating with nature to address social, economic and environmental challenges. *Environmental Research, 159*(November), 509–518. https://doi.org/10.1016/J.ENVRES.2017.08.032.

Fastenrath, S., Bush, J., & Coenen, L. (2020). Scaling-up nature-based solutions. Lessons from the Living Melbourne strategy. *Geoforum, 116*, 63–72. https://doi.org/10.1016/j.geoforum.2020.07.011.

Fedele, G., Donatti, C. I., Corwin, E., Pangilinan, M. J., Roberts, K., Lewins, M., ... Rambeloson, A. (2019). *Nature-based transformative adaptation: A practical handbook*. Arlington, VA, USA: Conservation International. https://doi.org/10.5281/zenodo.3386441.

Folke, C., Biggs, R., Norström, A. V., Reyers, B., & Rockström, J. (2016). Social-ecological resilience and biosphere-based sustainability science. *Ecology and Society*, *21*(3).

Frantzeskaki, N. (2019). Seven lessons for planning nature-based solutions in cities. *Environmental Science & Policy*, *93*, 101–111.

Frantzeskaki, N., Vandergert, P., Connop, S., Schipper, K., Zwierzchowska, I., Collier, M., & Lodder, M. (2020). Examining the policy needs for implementing nature-based solutions in cities: Findings from city-wide transdisciplinary experiences in Glasgow (UK), Ghent (Belgium) and Poznań (Poland). *Land Use Policy*, *96*. https://doi.org/10.1016/j.landusepol.2020.104688.

Hagedoorn, L. C., Koetse, M. J., van Beukering, P. J. H., & Brander, L. M. (2021). Reducing the finance gap for nature-based solutions with time contributions. *Ecosystem Services*, *52*, 101371. https://doi.org/10.1016/j.ecoser.2021.101371.

ILO, UNEP, & IUCN. (2022). *Decent work in nature-based solutions*. Retrieved from https://www.ilo.org/wcmsp5/groups/public/- - -ed_emp/documents/publication/wcms_863035.pdf.

IPCC. (2018). Annex I: Glossary [Matthews, J.B.R. (ed.)]. In V. Masson-Delmotte, P. Zhai, H.-O. Pörtner, D. Roberts, J. Skea, P. R. Shukla, … T. Waterfield (Eds.), *Global Warming of 1.5°C. An IPCC Special Report on the impacts of global warming of 1.5°C above pre-industrial levels and related global greenhouse gas emission pathways, in the context of strengthening the global response to the threat of climate change, sustainable development, and efforts to eradicate poverty* Cambridge University Press. https://www.ipcc.ch/sr15/chapter/glossary/.

Kooijman, E. D., McQuaid, S., Rhodes, M.-L., Collier, M. J., & Pilla, F. (2021). Innovating with nature: From nature-based solutions to nature-based enterprises. *Sustainability*, *13*(3), 1263. https://doi.org/10.3390/su13031263.

Kuhlicke, C., & Plavsic, J. (2021). *RECONECT's upscaling strategy*. Reconect Consortium. IUCN.

Lemos, M., & Agrawal, A. (2006). Environmental governance. *Annual Review of Environment and Resources*. https://doi.org/10.1146/annurev.energy.31.042605.135621.

Linnerooth-Bayer, J., Martin, J. C. G., Fresolone-Caparrós, A., Scolobig, A., & Aguilera Rodriguez, J. (2023). *Tackling policy barriers to nature-based solutions. Policy Brief of the PHUSICOS project, According to nature. Nature based solutions to reduce risks in mountain landscapes, EC H2020 Programme*. 4 pp https://pure.iiasa.ac.at/id/eprint/19133/1/PB39_Engage%20policy%20brief%5Bweb%5D.pdf.

Linnerooth-Bayer, J., Martin, J. C. G., Fresolone-Caparrós, A., Scolobig, A., Aguilera Rodriguez, J. J., Solheim, A., … Hoffstad Reutz, E. (2023). *Learning from NBS implementation barriers. Deliverable 5.4 of the PHUSICOS project, According to nature. Nature based solutions to reduce risks in mountain landscapes, EC H2020 Programme*. https://phusicos.eu/.

Lupp, G., & Zingraff-hamed, A. (2021). Nature-based solutions: Concept, evaluation, and governance. *Sustainability*, 1–5.

Mabon, L., Barkved, L., de Bruin, K., & Shih, W.-Y. (2022). Whose knowledge counts in nature-based solutions? Understanding epistemic justice for nature-based solutions through a multi-city comparison across Europe and Asia. *Environmental Science & Policy*, *136*, 652–664. https://doi.org/10.1016/j.envsci.2022.07.025.

Mačiulytė, E., & Durieux, E. (2020). *Public procurement of nature-based solutions: Addressing barriers to the procurement of urban NBS: Case studies and recommendations* (Independent Expert Report). European Union. Retrieved from https://data.europa.eu/doi/10.2777/561021.

Mahmoud, I. H., Morello, E., Rizzi, D., & Wilk, B. (2022). Localizing sustainable development goals (SDGs) through co-creation of nature-based solutions (NBS). In R. C. Brears (Ed.), *The Palgrave encyclopedia of urban and regional futures*. Cham: Palgrave Macmillan. https://doi.org/10.1007/978-3-030-87745-3_354.

Martín, E. G., Giordano, R., Pagano, A., Van Der Keur, P., & Costa, M. M. (2020). Using a system thinking approach to assess the contribution of nature based solutions to sustainable development goals. *Science of the Total Environment, 738*, 139693.

Martin, J. C. G., Irshaid, J., Linnerooth-Bayer, J., Scolobig, A., Aguilera Rodriguez, J. J., & Fresolone-Caparros, A. (2023). *Opportunities and barriers to NBS at the EU, national, regional and local scales, with suggested reforms and innovation. Deliverable 5.2 of the PHUSICOS project, According to Nature. Nature based solutions to reduce risks in mountain landscapes, EC H2020 Programme*. PHUSICOS. 58 pp https://phusicos.eu/.

Martin, J. C. G., Linnerooth-Bayer, J., Liu, W., Scolobig, A., & Balsiger, J. (2019). *Nature based solutions in-depth case study analysis of the characteristics of successful governance models, Deliverable 5.1 of the PHUSICOS project, According to nature. Nature-based solutions to reduce risks in mountain landscapes, EC H2020 Programme*. https://phusicos.eu/.

Martin, J.G.C., Linnerooth-Bayer, J., Scolobig, A., & Oen, A., et al. (forthcoming). Tackling the nature-based solution implementation gap: A review of barriers and enablers. Environmental Science and Policy.

Martin, J. G. C., Scolobig, A., Linnerooth-Bayer, J., Liu, W., & Balsiger, J. (2021). Catalyzing innovation: Governance enablers of nature-based solutions. *Sustainability, 13*(4), 1971.

Mayor, B., Toxopeus, H., McQuaid, S., Croci, E., Lucchitta, B., Reddy, S. E., … López Gunn, E. (2021). State of the art and latest advances in exploring business models for nature-based solutions. *Sustainability, 13*(13), 13. https://doi.org/10.3390/su13137413.

McQuaid, S. (2020). *Financing and business models guidebook*. Connecting Nature. Retrieved from https://connectingnature.oppla.eu/sites/default/files/uploads/finance-1.pdf.

McQuaid, S., Connop, S., & Müller, J. (2020). *Nature-based enterprises guidebook*. Connecting Nature. Retrieved from https://connectingnature.eu/sites/default/files/images/inline/Enterprise.pdf.

McQuaid, S., Kooijman, E. D., Rhodes, M.-L., & Cannon, S. M. (2021). Innovating with nature: Factors influencing the success of nature-based enterprises. *Sustainability, 13*(22), 12488. https://doi.org/10.3390/su132212488.

Naumann, S., & Davis, M. (2020). *Biodiversity and nature-based solutions—Analysis of EU-funded projects*. European Commission. https://doi.org/10.2777/183298.

O'Leary, B. C., Fonseca, C., Cornet, C. C., de Vries, M. B., Degia, A. K., Failler, P., … Roberts, C. M. (2022). Embracing nature-based solutions to promote resilient marine and coastal ecosystems. *Nature-Based Solutions*, 100044.

Ozment, S., Ellison, G., & Jongman, B. (2019). *Nature-based solutions for disaster risk management: Booklet*. Washington, DC, USA: World Bank Group. Retrieved from http://documents.worldbank.org/curated/en/253401551126252092/Booklet. Retrieved from.

Palomo, I., Locatelli, B., Otero, I., Colloff, M., Crouzat, E., Cuni-Sanchez, A., … Lavorel, S. (2021). Assessing nature-based solutions for transformative change. *One Earth, 4*(5), 730–741.

Ruangpan, L., Vojinovic, Z., Di Sabatino, S., Leo, L. S., Capobianco, V., Oen, A. M. P., … Lopez-Gunn, E. (2020). Nature-based solutions for hydro-meteorological risk reduction: A state-of-the-art review of the research area. *Natural Hazards and Earth System Sciences, 20*(1), 243–270. https://doi.org/10.5194/nhess-20-243-2020.

Runhaar, H., Wilk, B., Persson, Å., Uittenbroek, C., & Wamsler, C. (2018). Mainstreaming climate adaptation: Taking stock about "what works" from empirical research worldwide. *Regional Environmental Change, 18*, 1201–1210.

Scharlemann, J. P., Brock, R. C., Balfour, N., Brown, C., Burgess, N. D., Guth, M. K., … Kapos, V. (2020). Towards understanding interactions between Sustainable Development Goals: The role of environment–human linkages. *Sustainability Science, 15*, 1573–1584.

Schröter, B., Hack, J., Hüesker, F., Kuhlicke, C., & Albert, C. (2022). Beyond demonstrators— Tackling fundamental problems in amplifying nature-based solutions for the post-COVID-19 world. *NPJ Urban Sustainability*, *2*(1), 1. https://doi.org/10.1038/s42949-022-00047-z.

Scolobig, A., Aguilera, R. J., Linnerooth-Bayer, J., Martin, J. C. G., & Fresolone-Caparrós, A. (2023). *Policy and finance innovation for nature-based solutions. Policy Brief of the PHUSICOS project, According to nature. Nature based solutions to reduce risks in mountain landscapes, EC H2020 Programme*. 4 pp https://pure.iiasa.ac.at/id/eprint/19115/1/PB38_web.pdf.

Scolobig, A., Linnerooth Bayer, J., Martin, J. C. G., Altamirano, M., Duff, S., Del Seppia, N., ... Garcia, E. (2021). *The role of public and private sectors in mainstreaming nature-based solutions, second policy business forum synthesis, PHUSICOS project, According to nature. Nature based solutions to reduce risks in mountain landscapes, EC H2020 Programme*. Retrieved from https://phusicos.eu/.

Scolobig, A., Martin, J. C. G., Linnerooth-Bayer, J., Aguilera Rodriguez, J., Balsiger, J., Del Seppia, N., ... Zingraff-Hamed, A. (2023). *Governance innovation for the design, financing and implementation of NBS, and their application to the concept and demonstration projects, Deliverable 5.3 of the PHUSICOS project, According to Nature. Nature based solutions to reduce risks in mountain landscapes, EC H2020 Programme*. 104 pp. https://phusicos.eu/.

Scolobig, A., Martin, J. C. G., Linnerooth-Bayer, J., Balsiger, J., Andrea, A., Buckle, E., Calliari, E., et al. (2020). *Policy innovation for nature-based solutions in the disaster risk reduction sector. First policy business forum synthesis—PHUSICOS project. First policy business forum synthesis. PHUSICOS project, According to nature. Nature based solutions to reduce risks in mountain landscapes, EC H2020 Programme*. Retrieved from https://phusicos.eu/.

Seddon, N. (2022). Harnessing the potential of nature-based solutions for mitigating and adapting to climate change. *Science*, *376*(6600), 1410–1416.

Seddon, N., Chausson, A., Berry, P., Girardin, C. A. J., Smith, A., & Turner, B. (2020). Understanding the value and limits of nature-based solutions to climate change and other global challenges. *Philosophical Transactions of the Royal Society, B: Biological Sciences*, *375*(1794). https://doi.org/10.1098/rstb.2019.0120.

Silverman, D. (2010). *Qualitative research*. London: Sage.

Solheim, A., Capobianco, V., Oen, A., Kalsnes, B., et al. (2021). Implementing nature-based solutions in rural landscapes: Barriers experienced in the PHUSICOS project. *Sustainability*, *13*(3), 1461.

Steurer, R. (2013). Disentangling governance: A synoptic view of regulation by government, business and civil society. *Policy Sciences*, *46*. https://doi.org/10.1007/s11077-013-9177-y.

Surminski, S., Barnes, J., & Vincent, K. (2022). Can insurance catalyse government planning on climate? Emergent evidence from sub-Saharan Africa. *World Development*, *153*, 105830. https://doi.org/10.1016/j.worlddev.2022.105830.

Tilt, J. H., & Ries, P. D. (2021). Constraints and catalysts influencing green infrastructure projects: A study of small communities in Oregon (USA). *Urban Forestry & Urban Greening*, *63*, 127138. https://doi.org/10.1016/j.ufug.2021.127138.

Toxopeus, H., & Polzin, F. (2021). Reviewing financing barriers and strategies for urban nature-based solutions. *Journal of Environmental Management*. https://doi.org/10.1016/j.jenvman.2021.112371.

Trémolet, S., Kampa, E., Lago, M., Anzaldúa, G., Vidurre, R., Tarpey, J., ... Makropoulos, C. (2019). *Investing in nature for Europe water security*. The Nature Conservancy, Ecologic Institute and ICLEI.

United Nations Environment Assembly. (2022). *Resolution adopted by the United Nations Environment Assembly on 2 March 2022—Nature-based solutions for supporting sustainable development, UNEP/EA.5/Res.5*. Retrieved from https://wedocs.unep.org/bitstream/handle/20. 500.11822/39752/K2200677%20-%20UNEP-EA.5-Res.5%20-%20Advance.pdf?sequence= 1&isAllowed=y.

United Nations Environment Programme. (2021). *State of finance for nature 2021. Tripling investments in nature-based solutions by 2030*. United Nations Environment Programme.

United Nations Environment Programme (UNEP). (2022a). *Adaptation gap report 2022. Nairobi, Kenya*. Retrieved from https://www.unep.org/resources/adaptation-gap-report-2022?gclid= CjwKCAiAl9efBhAkEiwA4Torir-rP_xYyvAYMweE1zQrGysVOVIbU_ 66dVE62EqBgvkyBcrLKWofgxoC7MsQAvD_BwE.

United Nations Environment Programme (UNEP). (2022b). *State of Finance for Nature. Time to act: Doubling investment by 2025 and eliminating nature-negative finance flows*. Nairobi, Kenya: UNEP.

Vojinovic, Z. (2020). *Nature-based solutions for flood mitigation and coastal resilience. Analysis of EU-Funded Projects*. European Commission.

White House Council on Environmental Quality, & White House Office of Science and Technology Policy. (2022). *Opportunities to accelerate nature-based solutions*. [Report to the National Climate Task Force]. Retrieved from https://www.whitehouse.gov/wp-content/uploads/2022/11/ Nature-Based-Solutions-Roadmap.pdf.

World Business Council for Sustainable Development. (2019). Natural climate solutions: The business perspective. In *World Business Council for Sustainable Development*. Retrieved from https://docs.wbcsd.org/2019/09/WBCSD-Natural_climate_solutions-the_business_ perspective.pdf.

WWF. (2020). *Nature in all goals 2020*. Retrieved from https://wwfeu.awsassets.panda.org/ downloads/nature_in_all_goals_2020.pdf.

Zari, M. P., Kiddle, G. L., Blaschke, P., Gawler, S., & Loubser, D. (2019). Utilising nature-based solutions to increase resilience in Pacific Ocean Cities. *Ecosystem Services, 38*, 100968.

Zingraff-Hamed, A., Hüesker, F., Lupp, G., Begg, C., Huang, J., Oen, A., … Pauleit, S. (2020). Stakeholder mapping to co-create nature-based solutions: Who is on board? *Sustainability, 12*(20), 20. https://doi.org/10.3390/su12208625.

Chapter 2.2

NbS and policy communication

Cong Cong[a] and Diego Temkin[b]

[a]Urban Science and Planning, Massachusetts Institute of Technology, Cambridge, MA, United States, [b]Department of Urban Studies and Planning, MIT, Cambridge, MA, United States

1 Introduction

In the pursuit of solutions to anthropogenic climate change, nature-based solutions (NbS) interventions that leverage the benefits of human-scale nature offer a unique pathway toward building healthy and resilient communities. NbS are acknowledged as sustainable and cost-effective ways to nudge pro-environmental behaviors, thereby reducing urban greenhouse gas emissions (Keith et al., 2021; Ki & Lee, 2021; Pan et al., 2023). For instance, investing in trees along mobility corridors encourages people to bike on shaded, protected paths. Similarly, incorporating green facades on buildings helps regulate temperatures, cutting down energy consumption by reducing the need for air conditioning.

Stakeholder engagement is particularly important in NbS planning. Many NbS benefits are typically local, affecting only specific areas (Hansen & Pauleit, 2014). Consequently, the placement of NbS plays an essential role in defining who and what will benefit (Meerow & Newell, 2017). The Global Standard for NbS released by the International Union for Conservation of Nature (IUCN) titles one of their criteria as "NbS are based on inclusive, transparent, and empowering governance processes" (IUCN, 2020), emphasizing the significant role of stakeholders in successful NbS implementation, particularly in relation to public acceptance.

Employing suitable tools to facilitate dialog about the potential impacts of NbS can help align differing perspectives and promote social acceptance (Santoro et al., 2019). Important steps have already been taken in this direction. Guides have been developed to promote cross-sectoral coordination and collaboration (European Commission, 2015). Public participation support (PSS) tools invite community input in co-creating NbS plans, ensuring meaningful involvement and empowerment throughout the NbS process (Nesshöver et al., 2017; Sarabi et al., 2022; Wyborn & Bixler, 2013). Additionally, focus groups, workshops, and consensus conferences serve as deliberative forums for public learning and engagement (McEvoy et al., 2018; Page et al., 2020).

Nature-Based Solutions in Supporting Sustainable Development Goals
https://doi.org/10.1016/B978-0-443-21782-1.00008-7

Advances in interactive visualization using information and communication technology (ICT) are considered transformative for communicating complex issues (Wibeck et al., 2013). These tools are designed not only for small groups participating on-site but also for larger audiences. The focus of online communication has shifted from soliciting inputs toward addressing information deficits and building awareness, capacity, and agency on decision-making among the general public (Linders, 2013; Lumley et al., 2022). Interactive visualizations, in particular, are increasingly aimed at nonscientists to help bridge the digital divide in the information age (Wibeck et al., 2013).

The Nature-Based Solutions Dashboard developed by the research team is an interactive website that serves as a portal for exploring how we can collaborate with nature to find solutions to anthropogenic climate change. This dashboard is designed based on existing literature in scientific communication and stakeholder engagement (Malczewski, 2006; Pagano et al., 2019). It aims to enhance the collaborative planning process by providing a platform where NbS information is more accessible, relatable, and actionable and where various stakeholders can share their knowledge and preferences on NbS planning.

This chapter describes the development of the tool, together with its application in our research case of 50 EU cities. Section 2 provides a literature review on policy communication related to NbS planning, which informed the development of the dashboard. Section 3 offers a detailed description of the dashboard's content and format. Section 4 covers its application in our NbS research on EU cities, with discussion and conclusion in Sections 5 and 6.

2   Background

2.1   Policy communication needs for NbS

2.1.1   NbS for carbon emission mitigation

Many studies provide evidence that NbS can efficiently reduce urban carbon emissions. Following established definitions, urban NbS include waste and wastewater green infrastructure (GI), flood prevention GI, building green roofs and facades, street trees, green pavements, urban parks, bioremediation, preserved habitats, and urban agriculture (Babí Almenar et al., 2021; Castellar et al., 2021; Xie & Bulkeley, 2020). Anderson and Gough (2020) observed an average carbon dioxide reduction of 6% resulting from the application of GI. According to Ren et al. (2019), the average annual increase in carbon storage by urban GI offsets 3.9% of the increase in urban carbon emissions in China. Tomalty (2012) found that forestland, wetlands, and agricultural land in Ontario's Greenbelt around Toronto can annually store 86.6 million tons of carbon and sequester 200,000 tons of carbon.

NbS can reduce carbon emission levels through both direct and indirect pathways. Direct pathways typically refer to the natural growth of vegetation, during which they remove atmospheric CO_2 and store it in their biomass

(Nowak & Crane, 2002). Zhao et al. (2010) reported carbon sequestration by urban forests in Hangzhou, China, of over 1.3 MMT C/year, offsetting 18.57% of annual industrial carbon emissions with carbon storage equivalent to 1.75 times the annual industrial amount emitted. Indirect pathways reduce carbon emissions by nudging users to prioritize eco-friendly behaviors, thereby avoiding anthropogenic carbon output. For instance, Kuronuma et al.'s (2018) results showed significant energy savings for the experimental green roof, and the CO_2 emissions reduction rate ranges between 1.703 and 1.889 kg-CO_2/m/year. Lindsay et al. (2011) indicated the effect of micro-scale features on the streets, such as amenities and esthetics, in promoting increased walking, biking, and other pro-environment behaviors.

The advantages of implementing a diverse portfolio of ecosystem services to tackle carbon emission challenges are gaining recognition. Majidi et al. (2019) assessed the potential for implementing four different NbS (green roofs, pervious pavements, bio-retention cells, and rain gardens) and the benefit in reduction of flood risk and heat stress, but only at the small scale of a neighborhood in Bangkok. Understanding how the benefits of NbS are influenced by the urban structure of the cities at a large scale, along with identifying the co-benefits of various NbS types, is crucial for determining the most effective combinations to address climate change challenges in diverse urban contexts and built environments. This remains an ongoing topic requiring further research.

2.1.2   Communication toward stakeholder engagement

The broad implementation of NbS encounters documented barriers, as highlighted in the literature, such as uncertainties and limited information regarding their long-term behavior, as well as challenges in quantifying their multifaceted impacts (Pagano et al., 2019). Raymond et al. (2017) outlined requirements for each phase of the NbS planning cycle and proposed a seven-step policy process for their implementation. Frantzeskaki et al. (2020) viewed "policy needs" as disparities between current and desired outcomes at strategic, procedural, and operational tiers, systematically exploring the prerequisites for each phase of the NbS planning cycle. Both emphasized the crucial role of engaging citizens and businesses to collaboratively craft narratives, comprehend contextualized problem frameworks, and co-design NbS solutions.

Stakeholder engagement in urban planning has garnered significant attention due to its multifaceted benefits for both governmental bodies and wider social actors (Roberts, 2004). Laurian (2009) describes a desirable stakeholder engagement process in which the public collectively shapes planning outcomes while increasing their levels of social and political empowerment. The expectations of stakeholder participation in urban planning extend beyond merely assembling people. Effective participation aims to elevate residents' motivation, information, and awareness about local planning issues and processes, thereby transcending tokenism and moving toward genuine partnership, delegation, and citizen control. This concept aligns with Arnstein's (1969) "Ladder of Citizen Participation," which categorizes different levels of citizen involvement in decision-making.

Policy communication of NbS should foster meaningful participation. Rather than a one-way flow of information from government to citizens or vice versa, it should involve multidimensional interactions (Linders, 2013; Sieber & Johnson, 2015). This approach integrates decision-making, governance, organizational power, and citizen engagement, leading to co-evolution in collaborative urban planning (Innes & Booher, 2004). Importantly, the identities and voices of the poor, marginalized, and indigenous populations, often overlooked in mainstream political discourse, are increasingly being acknowledged and respected to promote political democracy and social justice (Friedmann, 1987; Young, 1990).

Regarding this communication strategy and drawing from the practice of PSS (Frantzeskaki et al., 2020), we propose that effective communication for developing an NbS implementation strategy should involve several steps. First, educate stakeholders about NbS and their benefits. Next, seek input from stakeholders on their preferences, including desired activities and management methods. Finally, involve stakeholders in an iterative process that allows for interaction and revision at different planning stages, aimed at prioritizing sites and solutions (Giordano et al., 2020; Kiss et al., 2022; Sarabi et al., 2022).

We respond to the needs for policy communication of NbS at the local level, focusing on identifying opportunities for shared knowledge and awareness building among planners and the public to enhance community impact. Most climate change studies concentrate on global and national levels, providing limited guidance for community-level policy communication (Kauark-Fontes et al., 2023; Sheppard et al., 2011). While promoting education, raising awareness, and engaging stakeholders are widely acknowledged as crucial, there is a need for specific examples and strategies tailored to local challenges, beyond generic recommendations (Sheppard et al., 2011). We advocate for the use of ICT-based visualization tools to exemplify our communication framework. The potential for affordable, advanced ICT-based visualization devices in science and policy communication is rapidly growing (Wibeck et al., 2013). Consequently, we see broader applications for these tools in other areas of sustainability communication.

2.2 Technology-enabled policy communication

The prevailing use of digital data and mobile devices has created more channels for soliciting inputs, thus diversifying ways for people to participate in decision-making processes. Various terms describe these new forms of participation in the information age, including electronic participation (eParticipation) (Macintosh, 2004; Tambouris et al., 2007), ICT-based participation (Palen &

Liu, 2007), technology-enabled participation (Desouza & Bhagwatwar, 2014), e-government (Yildiz, 2007), and smart governance (Tomor et al., 2019). Despite the different terminologies, there is a consensus that information, data, and communication technologies should aim to reach a broader audience and achieve deeper engagement (Macintosh, 2004; OECD, 2003; Tambouris et al., 2007). These technologies have removed time and space constraints for participation, making it easier to collect information and transform interested passersby into engaged participants (Höffken & Streich, 2013).

The roles of citizens in the participation process and their relationships with local governments and third parties are also expected to evolve accordingly. Based on Castells (1997), in a network society, network power can emerge, fostering shared heuristics, knowledge, and meanings among citizens. With greater autonomy in accessing and producing information, citizens are expected to act as partners rather than customers in public service delivery. Meanwhile, local governments are increasingly taking on new roles in convening and enabling civic actions (Linders, 2012). In the meantime, the reliance on data and technology enhances the role of intermediaries—including profit and nonprofit organizations, civic groups, and engaged citizens—in bridging the technical gap between citizens and the government (Pollock, 2011). These interactions among citizens, intermediaries, governments, and other actors conceptualize governance as a multiactor collaboration, where knowledge co-production becomes more relevant and viable with technological advances (Johnston & Hansen, 2011). This shift provides planners with an opportunity to rethink policy communication, networks, and institutional capacity within a technology-enabled collaborative framework, promoting increased citizen empowerment and inclusive decision-making in public governance.

The success of policy communications varies, with some being effective and others not. A broad framework for this includes the development of collaboration, particularly whether authorities are open and inclusive to public input (Attard et al., 2015; Linders, 2012). In the context of tech-based participation, while there are positive reports of digital platforms and technology-enabled participation improving problem-solving (Hong, 2015; Scott, 2015), there are also instances where no improvement is observed or where the use of technology and data exacerbates the digital divide (Lee & Kim, 2014; Neirotti et al., 2014). Understanding why this happens involves examining the conditions that favor successful outcomes.

2.3 Opportunities for successful policy communication

This section examines the generalizable conditions under which existing policy communication practices yield positive stakeholder engagement outcomes. Our goal is to provide a realistic perspective on the value of technology-based policy communication and help cities cultivate these favorable conditions. By reviewing prior literature that documents both the positive and negative outcomes of technology-based communication in local governance, we propose the following determining factors for success and identify opportunities for our subsequent research.

2.3.1   Government commitment to openness and cooperation

The degree to which local governments actively seek feedback from citizens and collaborate with private stakeholders significantly affects how users respond to participation opportunities (Dawes et al., 2016). Citizens often worry that communication systems won't lead to real changes (Kingston, 2007; Weerakkody et al., 2017). A study by Royo and Yetano (2015) highlights the varied reactions to government-led online surveys in two Spanish cities. It points out that technology-based engagement approaches will be ineffective if the government (1) focuses narrowly on disseminating information rather than genuinely promoting active citizen participation and (2) fails to take action or provide responses based on the participation received.

To initiate an open and collaborative relationship, governments need to showcase the results of stakeholder input. Demonstrating that citizen feedback genuinely influences planning outcomes is essential for building trust and encouraging continued participation, ensuring that stakeholders understand their involvement will lead to meaningful change rather than merely reinforcing existing processes (Janssen et al., 2012; Mergel, 2012).

Clear and accessible information is crucial when communicating with stakeholders. Studies on PSS indicate that overly complicated software and a lack of technical support discourage public involvement and undermine trust. Matheus et al. (2018) evaluated government data dashboards and argued that data visualization can either enhance or diminish transparency and trust, depending on whether mechanisms are in place to help the public interpret and understand the data. Lee and Kim (2014) noted that active participants are usually better educated, more affluent, and technologically savvy. Facilitating public engagement without widening the digital divide remains a significant challenge for democratic governance in the information age, and this issue has not been sufficiently addressed.

2.3.2   Stakeholders' intrinsic motivation and intangible rewards

Research shows that residents are more likely to participate in decision-making when the decisions directly affect their lives, serve a public cause (like the environment or education), or benefit their community (Kurniawan & de Vries, 2015; Lybeck, 2018; Mergel, 2012). For instance, Meijer and Potjer (2018) found that citizens are motivated to generate data when they align with their values and interests. Royo and Yetano (2015) summarized that citizens engage because they view the task as a worthwhile challenge (intrinsic motivation) or gain satisfaction from performing morally good actions (intangible rewards).

Intrinsic motivation and intangible rewards can be enhanced by user-friendly technology (Royo & Yetano, 2015). Traditional participation settings often limit time for public discussion and may not provide all available options for consideration (Brömmelstroet, 2017; Goodspeed, 2016). In contrast, online planning tools, crowdsourced projects, hackathons, and mobile participation devices are most effective when they empower participants to take action on public issues (Dawes et al., 2016). In the ImproveTheNeighborhood and See-ClickFix cases, citizens report street issues like surface cracks, flooding, and unattended waste disposal and track the progress of their reports through a full feedback cycle (Mergel, 2012). The focus on addressing community needs directly in public services significantly boosts public engagement in these initiatives.

Communicating NbS to the public requires aligning initiatives with community interests rather than solely prioritizing government agendas (Hollands, 2008; McCann, 2015). Involving local residents in selecting and designing NbS projects based on their specific needs and concerns can significantly enhance community buy-in and support (Lybeck, 2018; Meijer & Potjer, 2018). Highlighting the meaningful, positive changes residents can make in their environment fosters a sense of pride and satisfaction. Success stories and personal testimonials serve to illustrate the fulfillment and moral rewards associated with these projects (Kurniawan & de Vries, 2015; Mergel, 2012; Weerakkody et al., 2017). This approach underscores the community's collective impact, emphasizing the intrinsic value and intangible benefits of their involvement in NbS initiatives.

2.3.3 Systematic approaches and shared knowledge

Recent NbS studies advocate a shift from a solutionist to a systemic approach in urban development. It highlights the importance of understanding the interactions among nature, governance, and communities in shaping social-ecological problems and solutions (Mercado et al., 2024). Maintaining a balance between these three dimensions requires consensus on acknowledging the value of NbS within urban governance, preserving culturally diverse ways of relating to nature, and adopting new practices for long-term collaboration (Sarkki et al., 2024).

At the same time, nature-based interventions should be viewed within broader social contexts and political struggles such as environmental justice, social equity, and community resilience toward natural disasters, especially for marginalized communities (Anderson & Renaud, 2021). This entails reconceptualizing NbS not merely as isolated projects or technical components of green growth-focused action plans but as strategic interventions within broader socio-ecological transformation processes (Cong et al., 2023). By nurturing this comprehensive outlook, NbS could prompt broader inquiries into the criteria used to evaluate interventions within just transition strategies (Goličnik Marušić et al., 2023).

In the realm of systematic examinations of NbS, Mercado et al. (2024) highlighted the shift toward more reciprocal human-nature relationships, inclusive governance, and sustainable planning. Case studies focusing on the nexus of NbS implementation, particularly in small- to medium-sized cities and those in the Global South, are proposed to inform sustainable urban futures. Kauark-Fontes et al. (2023) demonstrated the increasing potential for integrating, institutionalizing, and developing NbS within urban governance, but limited shared knowledge and resource segregation remain barriers for NbS integration into urban policies and planning.

Effective policy communication can facilitate the exchange of insights in this interdependent relationship and align stakeholders toward common objectives (Frantzeskaki et al., 2020). For instance, clear and accessible policy briefs can inform stakeholders about the goals and strategies of nature-based interventions, encouraging their active participation and support. Stakeholder meetings and workshops provide platforms for diverse perspectives to be heard, fostering dialog and consensus-building. Online platforms and portals enable continuous communication and feedback, ensuring stakeholders remain engaged throughout the policy-making process (Fathejalali, 2016).

3 The NbS dashboard approach

3.1 Overview

The research team aims to develop a web-based platform for integrating NbS benefits and spatial allocation scenarios for better policy communication. The objective of the NbS Dashboard is to support interactive spatial planning processes at the local level, especially with regard to identifying implementation opportunities in the early phases of the NbS planning process (Sarabi et al., 2022; Sheppard et al., 2011). Therefore, the tool aims to improve communication among the stakeholders involved, support interdisciplinary discussions, and be easy to use.

NbS Dashboard offers an educational component and an interactive component. The educational component is designed to inform people about what NbS is. This approach can be helpful for users who are unfamiliar with NbS and need introductory information before engaging with the interactive features. The interactive component offers tools to view, search for, and compare NbS spatial allocations and benefits.

In terms of the audience, the NbS Dashboard aims for both internal communication and external communication. Internal communication is defined as that between the disciplines involved, such as spatial planners, urban developers, and environmental specialists. External communication mainly concerns the communication between local authorities and interested parties or stakeholders (Carsjens & Ligtenberg, 2007).

3.2   Content

3.2.1   The educational component

The purpose of the educational component is to explain what NbS approaches are and why we need them. The concept of NbS is relatively new and not well understood by the public. Many people mistakenly believe it only involves trees or parks, leading to confusion about what qualifies as NbS (Seddon et al., 2021). Since NbS is a long-term strategy, people often question the need for investment. This uncertainty can lead to inaction, as individuals feel that the issues related to climate change are too overwhelming for them to tackle in their daily lives (Frantzeskaki et al., 2020; Raymond et al., 2017).

Studies have highlighted the effectiveness of using a storyline approach to communicate new messages (Dahlstrom, 2014; Sundin et al., 2018). This concept involves presenting environmental messages through a simplified structure that clearly conveys cause and effect, identifies key actors, and outlines preferred solutions. The storyline in the educational component guides users through what NbS is, why it is necessary, what NbS looks like in urban settings, and the various benefits it offers. This approach aims to demystify NbS and highlights its importance and practicality in everyday life.

Effectively communicating about NbS requires more than just attracting attention; it involves spurring engagement and empowering people to make informed decisions (Pagano et al., 2019). To achieve this, communicators need to focus on the local impacts of NbS, making it relevant to the audience on a personal level. Highlighting concrete action strategies that can be implemented locally is crucial (Sheppard et al., 2011). This can be done by visualizing local adaptation measures and highlighting practical solutions.

For instance, to illustrate a city designed with NbS, we generated an image that incorporates NbS approaches into an urban landscape. This includes visualizing what urban parks, urban forests, and greenbelts might look like and how they could enhance the cityscape. This helps users easily imagine a city built with NbS (Fig. 1).

To visually demonstrate the benefits of NbS, we used side-by-side images to compare designs with and without NbS. For example, to show how investing in trees and protected greenspaces along mobility corridors can encourage biking and walking by providing shaded paths, we displayed two street images side by side. This visual comparison helps users understand and feel the potential impact of such investments (Fig. 2).

3.2.2   The interactive component

The interactive component of the NbS Dashboard features a map that responds to user clicks by showing our spatial allocation results for NbS. These results were generated from our previous research project aimed at maximizing the benefits of NbS by identifying the demands, locations, and types of NbS interventions in a spatially explicit manner. For more details, see Pan et al. (2023).

Vision a city designed with nature

Green Spaces
Mandate green space allocation in city planning, create permeable surfaces, convert vacant lots to community gardens, parks and green areas.

FIG. 1 Illustration of NbS approaches in cities.

Leverage NBS to Achieve Carbon-Neutral Cities

Nudge pro-environmental behaviors

Investing in trees and protected greenspaces along mobility corridors can offer a behavioral incentive, encouraging individuals to choose biking and walking along shaded, protected paths over driving cars on congested roads, thus contributing to carbon reduction efforts

Improve energy use

Green roof temperatures can be 30-40°F lower than those of conventional roofs. Using vegetated facades in built environments lowers air conditioning needs, reducing energy consumption and associated carbon emissions from power plants

Mitigate urban heat island effect

Green spaces mitigate carbon emissions by reducing urban heat levels through decreased sunlight absorption. They also store carbon in plants and aquatic environments, aiding in carbon sequestration. This dual effect helps offset emissions and mitigate climate change impacts in urban environments

FIG. 2 Side-by-side illustration of NbS benefits.

The results include a 30 by 30-m raster map showing the locations and sizes of five types of NbS: green infrastructures, green buildings, street trees, urban parks, and greenbelts. We also have quantified the emission reduction effect for each spatial unit.

This interactive map is designed to achieve the following four primary goals:

- Users can navigate to the city or place they are interested in and zoom in to their community.
- Users can see how NbS are spatially distributed in their neighborhood, including a breakdown of the five different types of NbS.
- Users can understand the emission reduction effects of each type of NbS and the overall impact.
- Users can offer feedback on where they think NbS should be prioritized.

We aim to show this information to emphasize the local impacts of NbS, making it personally relevant to the audience. Residents can explore and navigate the map to identify existing NbS opportunities in their area and discover potential NbS interventions they would like to see implemented. This localized focus helps individuals understand how NbS can directly benefit their communities (Fig. 3).

What needs to be noted is that the spatial allocation results presented here reflect outcomes from highly specialized modeling. This underscores the importance of motivating stakeholders to actively participate in the planning process to improve the validity of the system. Therefore, soliciting input is crucial, and these visualizations serve as an initial step in that direction.

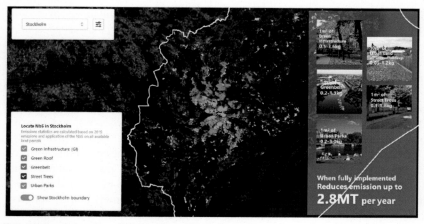

FIG. 3   Interactive map of NbS implementation effects.

3.3 Format

3.3.1 Features

Moser (2010) emphasizes that the choice of communication mediums and tools significantly impacts climate change communication. In the context of NbS visualization, we considered both the features selected and the visual representation based on our anticipated user profiles, expertise levels, and the intended information and tasks.

Communication tools range from highly interactive web-based platforms to static presentation formats displayed on screens or immersive environments. NbS Dashboard is a React application built on the JavaScript framework. In our educational component, we opted for light interactivity, such as flip cards and hover effects. This approach serves the following two purposes: first, interaction makes information more engaging and easier to comprehend. With this goal in mind, we curate information about NbS while avoiding overly academic language. Instead, we break it down into smaller, clickable pieces to encourage audience engagement. Second, we choose not to use excessive interactivity to prevent overwhelming viewers with information before they reach the second component, which is also the main focus of engagement.

For the interactive part, we utilized a highly responsive map. This decision was driven by the multilayered nature of the information. Decomposing the data and allowing viewers to explore areas of interest enhance their engagement and understanding. We offer a zoom-in and zoom-out feature along with various base maps to help users quickly and effectively locate themselves. The primary goal of this part is to invite feedback.

3.3.2 Visual representation

Visual representation encompasses various forms, including photographs, drawings, symbols, or maps, that are perceived by the human visual system (Evagorou et al., 2015). In our NbS visualization, we employ various geospatial representations to depict different types of NbS. We use landscape visualizations to illustrate how these elements integrate into the existing built environment or ecosystem.

For residents, effective communication with stakeholders requires presenting results in a way that does not overstress precision. Using sketch-like flippable icons, "read more" links, photos of familiar landscapes, and approximate benefit estimates will better capture their attention.

Planners, on the other hand, are more concerned with the cost and benefits of implementing NbS. While preparing this dashboard, we met a planner from Uppsala County in Sweden who had read our research and mentioned that having specific numbers "in black-and-white" would be extremely helpful for convincing politicians to allocate funding for NbS implementation. Consequently, we included specific numbers on the dashboard, showing how many kilograms of CO_2 per year could be removed or reduced from net emissions if this solution was implemented at all the shown locations.

Overall, we aim to use clear and accessible information to facilitate the sharing and exchange of knowledge between internal and external stakeholders. In determining the visualization format, we refer to earlier arguments about actively seeking feedback, providing relatable details, and engaging intrinsic values.

4   NbS integration for carbon emission reduction in European cities

The European Union (EU) has committed to decreasing net emissions by 57% by 2030, compared to 1990 levels, including land-use and carbon sequestration goals (European Parliament, 2022). This ambitious climate action plan involves 100 cities in EU member states that have pledged to achieve climate neutrality by 2030. The use of NbS aligns with EU climate policies and goals for addressing climate change. However, current policy programs do not adequately identify or leverage opportunities for NbS to mitigate carbon emissions beyond direct sequestration. Furthermore, limited urban land resources necessitate the optimal spatial allocation and configuration of NbS to maximize carbon emissions reductions.

In our team's previous study (see Pan et al., 2023), we assessed and quantified five potential carbon emissions reduction mechanisms for different types of NbS. Based on sector-wise carbon emissions data and the local context of 54 major European cities, we spatially allocated these 5 categories of NbS to each city. We estimated the emissions reduction potential for each sector and city, allocating one NbS implementation per land-use grid ($30\,m \times 30\,m$), which could function across different categories.

As a subsequent study and ongoing work, we visualized this information on an interactive NbS Dashboard to facilitate effective policy communication. Currently, our dashboard includes data for three EU cities. This tool enables the comparison of estimated emissions reduction potential against the 2030 climate neutrality targets of 54 cities, assessing the contributions of prioritized NbS toward these goals. Additionally, the dashboard serves as a foundation for developing more collaborative NbS plans.

There are two key tasks for future work. First, we aim to analyze the content of the storyline and how it is perceived by the audience, which is crucial for evaluating the impact of climate visualization and improving the dashboard. Second, we plan to conduct workshops with researchers, planners, and residents to adapt the model. We hope this dashboard will function as an effective policy communication tool, allowing users to interact and iterate through different planning stages. By going through this reviewing process, we aim to enhance its effectiveness for successful policy communication and implementation.

5 Discussion

As emphasized by Smith et al. (2005), educating residents about achieving environmental sustainability plays a crucial role in guiding cities toward carbon neutrality. The NbS Dashboard holds significant pedagogical potential in this regard. By leveraging the co-benefits related to social learning and cognitive resilience building (Colding & Barthel, 2013), cities can address the sociopsychological aspects of change, aligning with the calls made by Seyfang and Haxeltine (2012) for a shift in norm-setting environments in conjunction with broader societal shifts in values and beliefs (Westley et al., 2011).

Previous research on climate change communication underscores the importance of translating scientific uncertainties into terms understandable to nonscientific audiences (Lumley et al., 2022; Wibeck et al., 2013). However, it has also been argued that, for the general public, the most significant uncertainty lies not in scientific matters but rather in doubts regarding the effectiveness of various responses to climate change and the level of responsibility expected from individuals (Nisbet & Scheufele, 2009; Raymond et al., 2017). Determining who should bear responsibility for action and to what extent remains elusive in environmental discourse. The success of similar tools relies heavily on the willingness of participants in the planning process to engage on an equal footing during the initial stages.

The role of computation in informed decision-making has been a longstanding topic of discussion in planning, alongside the gap between the availability of planning support systems and their actual utilization in practice (Geertman, 2017; Jiang et al., 2020; Levenda et al., 2020; Vonk et al., 2005). This challenge is compounded by public concerns about the impacts of emerging computational tools such as the Generative AI (Yigitcanlar et al., 2021). In the context of carbon-neutral planning, valuable insights have been gleaned from data-driven research on eco-friendly and sustainable cities facilitated by IoT technologies.

An important consideration at this juncture is the readiness of planners to adopt and effectively utilize emerging technologies within the realm of carbon-neutral planning. This involves the following two key aspects: (1) the current and potential capacity of computational tools to meet planners' needs, and (2) planners' willingness and ability to embrace and leverage technology. Research indicates that barriers to the adoption of computational planning support systems in practice can be partially overcome through user involvement in technology development, offering valuable insights for AI-based planning tool developers. Ultimately, planners' awareness and understanding of AI will significantly influence the uptake of these technologies within the profession.

One methodological limitation of this study is that the measurement of fine-scale carbon emissions can be enhanced with further evidence if available. Land cover type data from Urban Atlas are useful but could be complemented with additional socioeconomic variables to better account for local-scale emissions. Also, our analysis only captures the directions of how NbS approaches affect carbon emissions, while the spatial distribution of NbS should be coordinated with other environmental and social benefits across social groups.

6 Conclusion

In this chapter, we discussed the significance of policy communication in the context of NbS planning. We have emphasized that successful policy communication hinges on stakeholders sharing knowledge, communicating on an equitable basis, and finding intrinsic motivation to participate. This pursuit extends beyond finding a suitable tool; it encompasses finding an effective means of communication. ICT-based tools that assist local authorities and planning offices, particularly in identifying options during the early stages of the planning process, hold promise from a communicative standpoint.

Subsequently, we introduced our initiative to develop an NbS Dashboard to visualize our NbS research findings as well as address our identified communication objectives. The case study in EU cities illustrates how the dashboard's information can contribute to the cities' climate policy goals and engage both internal and external stakeholders. We provided a detailed overview of the content and format of our dashboard, aligning with the communication principles we proposed. It is important to note that this dashboard serves as one proposed approach within the scope of our research, rather than a definitive solution. We look forward to further applications that align with the framework of successful policy communication and in accordance with planning ethics and values.

References

Anderson, V., & Gough, W. A. (2020). Evaluating the potential of nature-based solutions to reduce ozone, nitrogen dioxide, and carbon dioxide through a multi-type green infrastructure study in Ontario, Canada. *City and Environment Interactions*, 6, 100043.

Anderson, C. C., & Renaud, F. G. (2021). A review of public acceptance of nature-based solutions: The 'why', 'when', and 'how' of success for disaster risk reduction measures. *Ambio*, *50*(8), 1552–1573. https://doi.org/10.1007/s13280-021-01502-4.

Arnstein, S. (1969). A ladder of citizen participation. *Journal of the American Planning Association*, *35*(4), 216–224.

Attard, J., Orlandi, F., Scerri, S., & Auer, S. (2015). A systematic review of open government data initiatives. *Government Information Quarterly*, *32*(4), 399–418.

Babí Almenar, J., et al. (2021). Nexus between nature-based solutions, ecosystem services and urban challenges. *Land Use Policy*, *100*, 104898.

Brömmelstroet, M. (2017). Towards a pragmatic research agenda for the PSS domain. *Transportation Research Part A*, *104*, 77–83.

Carsjens, G. J., & Ligtenberg, A. (2007). A GIS-based support tool for sustainable spatial planning in metropolitan areas. *Landscape and Urban Planning*, *80*(1), 72–83. https://doi.org/10.1016/j.landurbplan.2006.06.004.

Castellar, J. A. C., et al. (2021). Nature-based solutions in the urban context: Terminology, classification and scoring for urban challenges and ecosystem services. *Science of the Total Environment*, *779*, 146237.

Castells, M. (1997). *The power of identity: The information age: Economy, society, and culture. Vol. 2.* Malden, MA: Blackwell.

Colding, J., & Barthel, S. (2013). The potential of 'urban green commons' in the resilience building of cities. *Ecological Economics, 86,* 156–166.

Cong, C., Pan, H., Page, J., et al. (2023). Modeling place-based nature-based solutions to promote urban carbon neutrality. *Ambio, 52,* 1297–1313. https://doi.org/10.1007/s13280-023-01872-x.

Dahlstrom, M. F. (2014). Using narratives and storytelling to communicate science with nonexpert audiences. *Proceedings of the National Academy of Sciences of the United States of America, 111*(Suppl 4), 13614–13620. https://doi.org/10.1073/pnas.1320645111.

Dawes, S. S., Vidiasova, L., & Parkhimovich, O. (2016). Planning and designing open government data programs: An ecosystem approach. *Government Information Quarterly, 33*(1), 15–27.

Desouza, K., & Bhagwatwar, A. (2014). Technology-enabled participatory platforms for civic engagement: The case of U.S. cities. *Journal of Urban Technology, 21*(4), 25–50.

European Commission. (2015). *Towards an EU research and innovation policy agenda for nature-based solutions & re-naturing cities.*

European Parliament. (2022). *Fit for 55: Parliament agrees to higher EU carbon sink ambitions by 2030.* https://www.europarl.europa.eu/news/en/press-room/20220603IPR32133/fit-for-55-parliament-agrees-to-higher-eu-carbon-sink-ambitions-by-2030 (Accessed 27 May 2024).

Evagorou, M., Erduran, S., & Mäntylä, T. (2015). The role of visual representations in scientific practices: From conceptual understanding and knowledge generation to 'seeing' how science works. *International Journal of STEM Education, 2,* 11. https://doi.org/10.1186/s40594-015-0024-x.

Fathejalali, A. (2016). *Enhancing citizen engagement in urban planning processes through mobile participation (mParticipation).* Technische Universität Berlin. Retrieved June 24, 2020 from https://depositonce.tu-berlin.de/bitstream/11303/6244/4/fathejalali_arman.pdf.

Frantzeskaki, N., Vandergert, P., Connop, S., Schipper, K., Zwierzchowska, I., Collier, M., et al. (2020). Examining the policy needs for implementing nature-based solutions in cities: Findings from city-wide transdisciplinary experiences in Glasgow (UK), Genk (Belgium) and Poznań (Poland). *Land Use Policy, 96,* 104688. https://doi.org/10.1016/j.landusepol.2020.104688.

Friedmann, J. (1987). *Planning in the public domain: From knowledge to action.* Princeton, NJ: Princeton University Press.

Geertman, S. (2017). PSS: Beyond the implementation gap. *Transportation Research Part A: Policy and Practice, 104,* 70–76.

Giordano, R., Pluchinotta, I., Pagano, A., Scrieciu, A., & Nanu, F. (2020). Enhancing nature-based solutions acceptance through stakeholders' engagement in co-benefits identification and trade-offs analysis. *Science of the Total Environment, 713,* 136552.

Goličnik Marušić, B., Dremel, M., & Ravnikar, Ž. (2023). A frame of understanding to better link nature-based solutions and urban planning. *Environmental Science & Policy, 146,* 47–56. https://doi.org/10.1016/j.envsci.2023.05.005.

Goodspeed, R. (2016). Digital knowledge technologies in planning practice: From black boxes to media for collaborative inquiry. *Planning Theory and Practice, 17*(4), 577–600.

Hansen, R., & Pauleit, S. (2014). From multifunctionality to multiple ecosystem services? A conceptual framework for multifunctionality in green infrastructure planning for urban areas. *Ambio, 43*(4), 516–529. https://doi.org/10.1007/s13280-014-0510-2.

Höffken, S., & Streich, B. (2013). Mobile participation. In C. N. Silva (Ed.), *Citizen E-participation in urban governance* (pp. 199–225). IGI Global.

Hollands, R. (2008). Will the real smart city please stand up? *City, 12*(3), 303–320.

Hong, S. (2015). Citizen participation in budgeting: A trade-off between knowledge and inclusiveness? *Public Administration Review, 75*(4), 572–582.

Innes, J. E., & Booher, D. E. (2004). Reframing public participation: Strategies for the 21st century. *Planning Theory & Practice, 5*(4), 419–436.

International Union for Conservation of Nature and Natural Resources (IUCN). (2020). *Global standard for nature-based solutions. A user-friendly framework for the verification, design and scaling up of NbS* (1st ed.). Gland, Switzerland: IUCN. https://doi.org/10.2305/IUCN.CH.2020.08.en.

Janssen, M., Charalabidis, Y., & Zuiderwijk, A. (2012). Benefits, adoption barriers and myths of open data and open government. *Information Systems Management, 29*(4), 258–268.

Jiang, H., Geertman, S., & Witte, P. (2020). Ignorance is bliss? An empirical analysis of the determinants of PSS usefulness in practice. *Computers, Environment and Urban Systems, 83*, 101505.

Johnston, E., & Hansen, D. (2011). Design lessons for smart governance infrastructures. In D. Ink, A. Balutis, & T. Buss (Eds.), *American governance 3.0: Rebooting the public square?* National Academy of Public Administration.

Kauark-Fontes, B., Marchetti, L., & Salbitano, F. (2023). Integration of nature-based solutions (NBS) in local policy and planning toward transformative change. Evidence from Barcelona, Lisbon, and Turin. *Ecology and Society, 28*(2). https://doi.org/10.5751/ES-14182-280225.

Keith, H., et al. (2021). Evaluating nature-based solutions for climate mitigation and conservation requires comprehensive carbon accounting. *Science of the Total Environment, 769*, 144341.

Ki, D., & Lee, S. (2021). Analyzing the effects of Green View Index of neighborhood streets on walking time using Google Street View and deep learning. *Landscape and Urban Planning, 205*, 103920.

Kingston, R. (2007). Public participation in local policy decision-making: The role of web-based mapping. *The Cartographic Journal, 44*(2), 138–144.

Kiss, B., Sekulova, F., Hörschelmann, K., Salk, C. F., Takahashi, W., & Wamsler, C. (2022). Citizen participation in the governance of nature-based solutions. *Environmental Policy and Governance, 32*(3), 247–272. https://doi.org/10.1002/eet.1987.

Kurniawan, M., & de Vries, W. (2015). The contradictory effects in efficiency and citizens' participation when employing geo-ICT apps within local government. *Local Government Studies, 41*(1), 119–136.

Kuronuma, T., Watanabe, H., Ishihara, T., Kou, D., Toushima, K., Ando, M., et al. (2018). CO_2 payoff of extensive green roofs with different vegetation species. *Sustainability, 10*, 2256.

Laurian, L. (2009). Trust in planning: Theoretical and practical considerations for participatory and deliberative planning. *Planning Theory & Practice, 10*(3), 369–391. https://doi.org/10.1080/14649350903229810.

Lee, J., & Kim, S. (2014). Active citizen E-participation in local governance: Do individual social capital and E-participation management matter? In *Paper presented at the 47th Hawaii international conference on system sciences (Hawaii, January 6–9, 2014)*.

Levenda, A., Rock, N., & Miller, B. (2020). Rethinking public participation in the smart city. *The Canadian Geographe [Le Géographe Canadien], 64*(3), 344–358.

Linders, D. (2012). From e-government to we-government: Defining a typology for citizen coproduction in the age of social media. *Government Information Quarterly, 29*(4), 446–454. https://doi.org/10.1016/j.giq.2012.06.003.

Linders, D. (2013). Towards open development: Leveraging open data to improve the planning and coordination of international aid. *Government Information Quarterly, 30*(4), 426–434.

Lindsay, G., Macmillan, A., & Woodward, A. (2011). Moving urban trips from cars to bicycles: Impact on health and emissions. *Australian and New Zealand Journal of Public Health, 35*, 54.

Lumley, S., Sieber, R., & Roth, R. (2022). A framework and comparative analysis of web-based climate change visualization tools. *Computers and Graphics*, *103*, 19–30. https://doi.org/10.1016/j.cag.2021.12.007.

Lybeck, R. (2018). Mobile participation in urban planning: Exploring a typology of engagement. *Planning Practice and Research*, *33*(5), 523–539.

Macintosh, A. (2004). Characterizing e-participation in policymaking. In *Paper presented at 37th annual international conference on system sciences, Big Island, HI, USA.*

Majidi, A. N., Vojinovic, Z., Alves, A., Weesakul, S., Sanchez, A., Boogaard, F., et al. (2019). Planning nature-based solutions for urban flood reduction and thermal comfort enhancement. *Sustainability*, *11*. https://doi.org/10.3390/su11226361.

Malczewski, J. (2006). GIS-based multicriteria decision analysis: A survey of the literature. *International Journal of Geographical Information Science*, *20*(7), 703–726. https://doi.org/10.1080/13658810600661508.

Matheus, R., Janssen, M., & Maheshwari, D. (2018). Data science empowering the public: Data-driven dashboards for transparent and accountable decision-making in smart cities. *Government Information Quarterly*, *37*(3), 101284. https://doi.org/10.1016/j.giq.2018.01.006.

McCann, L. (2015). *Experimental modes of civic engagement in civic tech*. Chicago: Smart Chicago Collaborative.

McEvoy, S., van de Ven, F. H. M., Blind, M. W., et al. (2018). Planning support tools and their effects in participatory urban adaptation workshops. *Journal of Environmental Management*, *207*, 319–333.

Meerow, S., & Newell, J. P. (2017). Spatial planning for multifunctional green infrastructure: Growing resilience in Detroit. *Landscape and Urban Planning*, *159*, 62–75. https://doi.org/10.1016/j.landurbplan.2016.10.005.

Meijer, A., & Potjer, S. (2018). Citizen-generated open data: An explorative analysis of 25 cities. *Government Information Quarterly*, *35*, 613–621.

Mercado, G., Wild, T., Hernandez-Garcia, J., Baptista, M. D., van Lierop, M., Bina, O., et al. (2024). Supporting nature-based solutions via nature-based thinking across European and Latin American cities. *Ambio*, *53*(1), 79–94. https://doi.org/10.1007/s13280-023-01920-6.

Mergel, I. (2012). *Distributed democracy: SeeClickFix.com for crowdsourced issue reporting.* Available at SSRN https://ssrn.com/abstract=1992968.

Moser, S. (2010). Communicating climate change: History, challenges, process and future directions. *Wiley Interdisciplinary Reviews: Climate Change*, *1*(1), 31–53. https://doi.org/10.1002/wcc.11.

Neirotti, P., De Marco, A., Cagliano, A., Mangano, G., & Scorrano, F. (2014). Current trends in Smart City initiatives: Some stylized facts. *Cities*, *38*, 25–36.

Nesshöver, C., Assmuth, T., Irvine, K. N., Rusch, G. M., Waylen, K. A., Delbaere, B., et al. (2017). The science, policy and practice of nature-based solutions: An interdisciplinary perspective. *The Science of the Total Environment*, *579*, 1215–1227.

Nisbet, M., & Scheufele, D. (2009). What's next for science communication? Promising directions and lingering distractions. *American Journal of Botany*, *96*(10), 1767–1778. https://doi.org/10.3732/ajb.0900041.

Nowak, D., & Crane, D. (2002). Carbon storage and sequestration by urban trees in the USA. *Environmental Pollution*, *116*, 381–389.

OECD. (2003). *The e-government imperative, OECD e-government studies*. Paris: OECD Publishing.

Pagano, A., Pluchinotta, I., Pengal, P., Cokan, B., & Giordano, R. (2019). Engaging stakeholders in the assessment of NBS effectiveness in flood risk reduction: A participatory System Dynamics

Model for benefits and co-benefits evaluation. *Science of the Total Environment, 690,* 543–555. https://doi.org/10.1016/j.scitotenv.2019.07.059.

Page, J., Mörtberg, U., Destouni, G., Ferreira, C., Näsström, H., & Kalantari, Z. (2020). Open-source planning support system for sustainable regional planning: A case study of Stockholm County, Sweden. *Environment and Planning B: Urban Analytics and City Science, 47*(8), 1508–1523. https://doi.org/10.1177/2399808320919769.

Palen, L., & Liu, S. (2007). Citizen communications in crisis: Anticipating a future of ICT-supported public participation. *Proceedings of the SIGCHI Conference on Human Factors in Computing Systems* (pp. 727–736). Association for Computing Machinery. https://doi.org/10.1145/1240624.1240736.

Pan, H., Page, J., Shi, R., et al. (2023). Contribution of prioritized urban nature-based solutions allocation to carbon neutrality. *Nature Climate Change, 13,* 862–870. https://doi.org/10.1038/s41558-023-01737-x.

Pollock, R. (2011 March 31). *Building the (Open) data ecosystem.* Retrieved from https://blog.okfn.org/2011/03/31/building-the-open-data-ecosystem/.

Raymond, C. M., Frantzeskaki, N., Kabisch, N., Berry, P., Breil, M., Nita, M. R., et al. (2017). A framework for assessing and implementing the co-benefits of nature-based solutions in urban areas. *Environmental Science & Policy, 77,* 15–24. https://doi.org/10.1016/j.envsci.2017.07.008.

Ren, Z., Zheng, H., He, X., Zhang, D., Shen, G., & Zhai, C. (2019). Changes in spatio-temporal patterns of urban forest and its above-ground carbon storage: Implication for urban CO_2 emissions mitigation under China's rapid urban expansion and greening. *Environment International, 129,* 438–450.

Roberts, N. (2004). Public deliberation in an age of direct citizen participation. *The American Review of Public Administration, 34*(4), 315–353.

Royo, S., & Yetano, A. (2015). "Crowdsourcing" as a tool for e-participation: Two experiences regarding CO2 emissions at municipal level. *Electronic Commerce Research, 15,* 323–348.

Santoro, S., Pluchinotta, I., Pagano, A., Pengal, P., Cokan, B., & Giordano, R. (2019). Assessing stakeholders' risk perception to promote nature based solutions as flood protection strategies: The case of the Glinščica river (Slovenia). *Science of the Total Environment, 655,* 188–201. https://doi.org/10.1016/j.scitotenv.2018.11.116.

Sarabi, S., Han, Q., de Vries, B., & Romme, A. (2022). The nature-based solutions planning support system: A playground for site and solution prioritization. *Sustainable Cities and Society, 78,* 103608.

Sarkki, S., Haanpää, O., Heikkinen, H. I., Hiedanpää, J., Kikuchi, K., & Räsänen, A. (2024). Mainstreaming nature-based solutions through five forms of scaling: Case of the Kiiminkijoki River basin, Finland. *Ambio, 53*(2), 212–226. https://doi.org/10.1007/s13280-023-01942-0.

Scott, T. (2015). Does collaboration make any difference? Linking collaborative governance to environmental outcomes. *Journal of Policy Analysis and Management, 34*(3), 537–566.

Seddon, N., Smith, A., Smith, P., Key, I., Chausson, A., et al. (2021). Getting the message right on nature-based solutions to climate change. *Global Change Biology, 27*(8), 1518–1546.

Seyfang, G., & Haxeltine, A. (2012). Growing grassroots innovations: Exploring the role of community-based initiatives in governing sustainable energy transitions. *Environment and Planning. C, Government & Policy, 30*(3), 381–400. https://doi.org/10.1068/c10222.

Sheppard, S., Shaw, A., Flanders, D., Burche, S., Wiek, A., Carmichael, J., et al. (2011). Future visioning of local climate change: A framework for community engagement and planning with scenarios and visualisation. *Futures, 43*(4), 400–412. https://doi.org/10.1016/j.futures.2011.01.009.

Sieber, R. E., & Johnson, P. A. (2015). Civic open data at a crossroads: Dominant models and current challenges. *Government Information Quarterly, 32*(3), 308–315.

Smith, A., Stirling, A., & Berkhout, F. (2005). The governance of sustainable socio-technical transitions. *Research Policy, 34*(10), 1491–1510.

Sundin, A., Andersson, K., & Watt, R. (2018). Rethinking communication: Integrating storytelling for increased stakeholder engagement in environmental evidence synthesis. *Environmental Evidence, 7*, 6. https://doi.org/10.1186/s13750-018-0116-4.

Tambouris, E., Macintosh, A., Coleman, S., Wimmer, M., Vedel, T., Westholm, H., et al. (2007). *Introducing eParticipation*. Retrieved June 24, 2020 from http://www.ifib-consult.de/publikationsdateien/Introducing_eParticipation_DEMO-net_booklet_1.pdf.

Tomalty, R. (2012). *Carbon in the Bank—Ontario's greenbelt and its role in mitigating climate change*. Vancouver: David Suzuki Foundation. https://davidsuzuki.org/wp-content/uploads/2012/08/carbon-bank-ontario-greenbelt-role-mitigating-climate-change.pdf (Accessed 29 December 2022).

Tomor, Z., Meijer, A., Michels, A., & Geertman, S. (2019). Smart governance for sustainable cities: Findings from a systematic literature review. *Journal of Urban Technology, 26*(4), 3–27.

Vonk, G., Geertman, S., & Schot, P. (2005). Bottlenecks blocking widespread usage of planning support systems. *Environment and Planning A: Economy and Space, 37*(5), 909–924.

Weerakkody, V., Irani, Z., Kapoor, K., Sivarajah, U., & Dwivedi, Y. (2017). Open data and its usability: An empirical view from the Citizen's perspective. *Information Systems Frontiers, 19*, 285–300. https://doi.org/10.1007/s10796-016-9679-1.

Westley, F., Olsson, P., Folke, C., et al. (2011). Tipping toward sustainability: Emerging pathways of transformation. *Ambio, 40*, 762–780. https://doi.org/10.1007/s13280-011-0186-9.

Wibeck, V., Neset, T., & Linnér, B. (2013). Communicating climate change through ICT-based visualization: Towards an analytical framework. *Sustainability, 5*, 4760–4777. https://doi.org/10.3390/su5114760.

Wyborn, C., & Bixler, R. P. (2013). Collaboration and nested environmental governance: Scale dependency, scale framing, and cross-scale interactions in collaborative conservation. *Journal of Environmental Management, 123*, 58–67. https://doi.org/10.1016/j.jenvman.2013.03.014.

Xie, L., & Bulkeley, H. (2020). Nature-based solutions for urban biodiversity governance. *Environmental Science and Policy, 110*, 77–87.

Yigitcanlar, T., Mehmood, R., & Corchado, J. M. (2021). Green artificial intelligence: Towards an efficient, sustainable and equitable technology for Smart Cities and futures. *Sustainability, 13* (16), 16. https://doi.org/10.3390/su13168952.

Yildiz, M. (2007). E-government research: Reviewing the literature, limitations, and ways forward. *Government Information Quarterly, 24*(3), 646–665.

Young, I. (1990). *Justice and the politics of difference*. Princeton, NJ: Princeton University Press.

Zhao, M., Kong, Z.-h., Escobedo, F. J., & Gao, J. (2010). Impacts of urban forests on offsetting carbon emissions from industrial energy use in Hangzhou, China. *Journal of Environmental Management, 91*(4), 807–813.

Chapter 2.3

The integration and adoption of the concept of urban resilience into policy in the Netherlands

Nina Escriva Fernandez[a] and Haozhi Pan[b]

[a]School of International and Public Affairs, Shanghai Jiao Tong University, Shanghai, China, [b]Shanghai Jiao Tong University, Shanghai, China

1 Introduction

The idea of nature-based solutions (NbS) and urban resilience has been widely adopted by major research and innovation projects in real-life laboratories (Zingraff-Hamed et al., 2021). However, the concept has not yet been intrinsically integrated into urban governance and policy agendas (Calliari et al., 2022), and they are still far from being mainstreamed in urban development (Dorst et al., 2022). One of the main barriers is that policymakers usually work with knowledge segregated "in silos" among departments, disciplines, sectors, and jurisdictions, often facing conflicting agendas (Sarabi et al., 2019). Siloing appears to be the status quo in most city governments, in which different departments and institutions operate based on distinct visions, ways of thinking, objectives, and legal structures (O'Donnell et al., 2018). In contrast, the inherent multifunctionality, multidisciplinarity, multiform, and place-based characteristics of NbS and urban resilience bring the need for cross-scale collaborative governance to efficiently achieve transformative change (Dorst et al., 2019).

Thus, NbS and urban resilience need to be connected beyond physical, jurisdictional, and temporal boundaries, requiring the interaction of multiple governance and policymaking levels, including urban, municipal, metropolitan, regional, and national scales, to allow a connection between the tactical and the strategic level development (Kabisch et al., 2022). Such connection represents a break of silos and a shift toward a more collective, flexible, and adaptive way of development to be integrated into urban governance, policy arrangements, and instruments (Dorst et al., 2022).

Nature-Based Solutions in Supporting Sustainable Development Goals
https://doi.org/10.1016/B978-0-443-21782-1.00009-9

Institutional incentives play an important role in policy adoption (Coaffee et al., 2018; Huck et al., 2020; Massey & Huitema, 2013). However, even when no institutional pressure exists, cities frequently choose to act and adopt a policy (Lee, 2013). Policy adoption and diffusion theory discusses the various internal and external motivators that can lead to policy adoption by a unit of governance. An example of an internal motivator for policy adoption is the stress that a city is subject to, such as flood risk. Policy diffusion refers to "policies in one unit (country, state, city, etc.) being influenced by the policies of other units" (Gilardi & Wasserfallen, 2019) and views units of governance as interdependent. An example of an external motivator that can lead to policy adoption is participation in a policy network.

As the resilience framework is not yet mainstream among cities (Huck et al., 2020) but can play an important role in creating a sustainable future (Urban Resilience Hub, n.d.), research into the drivers for cities to integrate the concept of resilience into their policy can help to make this type of policy more mainstream. Utilizing policy adoption and diffusion theory, this study will investigate the motivators for policy adoption by looking specifically at the integration of resilience as a concept by the 50 largest municipalities in the Netherlands. Thus, this chapter aims to examine the current policy diffusion mechanisms for urban resilience policies at the municipal level, finding its main drivers and identifying key areas for improvement, and herein seeks to answer the following questions: (1) What are the internal and external factors for policy diffusion? (2) What are the most important drivers of policy adoption: internal or external factors?

Sections 2.1–2.3 will answer the first question by presenting the literature review on the internal and external factors for policy diffusion. In Section 2.4, the case study on the Netherlands will be introduced and utilized to answer the second question.

2 Factors for policy diffusion

2.1 Internal factors for policy diffusion

Various factors motivate a unit of governance to adopt a policy. These motivators can be referred to as drivers of policy adoption. There are both internal and external factors driving policy decisions (Gordon, 2013; Kammerer & Namhata, 2018; Mallinson, 2021; Massey et al., 2014; Matisoff & Edwards, 2014). Internal determinants are factors within the governance unit such as political motivations, problem pressure, available resources, or regulations. These can be characterized as "political, economic, or social characteristics" that are internal to the unit of governance (Berry & Berry, 2007). These "contextual conditions" have been identified by some scholars as potentially the single most important factor resulting in policy adoption and innovation (Otto et al., 2021; Schoenefeld et al., 2022). These factors can be motivators for policy

TABLE 1 Internal/external category definitions.

Factor	Category definition/criteria	Source
Internal	Describes a characteristic or part of a municipality regardless of the city's level of control over this characteristic or part	Berry and Berry (2007), Massey et al. (2014), Otto et al. (2021), and Schoenefeld et al. (2022)
External	Describes interactions between different municipalities	Berry and Berry (2007), Kuhlmann et al. (2019), and Shipan and Volden (2008)

innovation as well as barriers to innovation (Mohr, 1969). These factors can include any type of characteristic of a municipality and do not refer to a municipality's agency or control over said characteristics. It is possible that this internal condition is outside the scope of influence of the municipality (see Table 1).

However, when municipalities adopt new policies, it is unlikely these policies have been developed without outside influence (Eraydin & Özatağan, 2021). The process in which governance units influence each other is known as policy diffusion. Policy diffusion describes "the spread of an object in space and time" (Kincaid, 2004). Gilardi and Wasserfallen (2019) define policy diffusion as "policies in one unit (country, state, city, etc.) being influenced by the policies of other units." Rogers (1983) defines diffusion as "the process by which an innovation is communicated through certain channels over time among the members of a social system." Policy diffusion is thus closely interlinked with policy innovation and adoption.

Policy diffusion focuses on interactions between entities (Berry & Berry, 2007) (see Section 2.4). The internal factors focus on a characteristic, or part (including individuals) of a municipality. These can be inside or outside the influence of the municipality (Massey et al., 2014).

Although there are numerous possible internal factors contributing to policy adoption and diffusion, the ones commonly discussed in the literature are as follows.

2.1.1   City size/resources

One of the most common factors scholars include in their research on both policy adoption and diffusion is city size. Salvia et al. (2021) find that larger cities have generally higher ambitions in terms of climate goals and thus are more likely to have adopted climate-related policies (Verschueren, 2022). Scholars have found that smaller cities are more likely to be incentivized by emulation or are more likely to "imitate" a policy if a big city near them adopted this policy before (Abel, 2021; Shipan & Volden, 2008).

One explanation for the role of city size is that the bigger the city, the more resources it has at its disposal (Abel, 2021). Human or financial resources can play an important role in either stimulating or inhibiting a municipality from adopting a policy (Aguiar et al., 2018; Araos et al., 2016; DellaVigna & Kim, 2022; Matisoff & Edwards, 2014; Otto et al., 2021; Ryan, 2015).

Resources are important not only for the execution of policy but also for the process leading up to policy execution. Municipal employees are an important resource in, for example, policy research and exploration, and smaller municipalities have fewer employees. Small- and medium-size cities also have less resources to spend on participating in networks that can stimulate policy innovation (Fünfgeld, 2014; Haupt et al., 2020; Häußler & Haupt, 2021). The degree to which resources play a role in policy adoption also depends on the policy and the amount of resources it requires (Ryan, 2015). Resilience policy is an interdisciplinary policy aiming at various sectors of society and can be expected to require many resources.

2.1.2 Problem pressure

If a municipality faces a certain issue or is at risk of something (such as flooding or a high unemployment rate), it is likely that policymakers are incentivized to resolve the issue or reduce the risk. This can be both from intrinsic motivation from policymakers and also because of constituent pressure. Problem severity has been found to have an effect on the likeliness of policy adoption; for example, the severity of air quality increases the likeliness of the adoption of climate policy aimed at mitigating this problem (Matisoff & Edwards, 2014). Problem-solving potential has also been found to play a role in the likeliness of policy adoption, for example, wind potential in regard to renewable energy policy (Matisoff & Edwards, 2014). It is also likely that certain geographic features of a city play a role in policy adoption, as they can increase the stress on a municipality. However, this is possibly only when this geographic feature brings them factually at higher risk. Whether a city is located on the coast or not has no significant results in adopting climate policy if it is not directly associated with higher risks or stresses (Lee, 2013; Rashidi & Patt, 2018).

2.1.3 Globalization

Scholars have investigated other factors regarding policy adoption including the level of globalization of a city (Lee & Koski, 2014). A city's level of globalization (in terms of economic and social interconnectedness) plays an important role in its decision to participate in global networks (Lee, 2013). These networks can assist and play a role in policy adoption. Heikkinen et al. (2020) found that wealthier countries are more likely to be part of climate networks. This can be due to resources, as cities need to have resources to be able to be part of a network (Lee, 2013). However, it can also result from wanting to join a community of similar countries as it shows dedication to the topic (Heikkinen et al., 2020;

Lee, 2013). The 100 Resilient Cities Network, for example, is deemed exclusive and has prerequisites in order to become part of the network (Zebrowski, 2020). Countries participating in climate networks are more likely to have adopted climate-related measures. However, whether this is causal or if the city joined the network because of an existing focus on climate adaptation is unclear: "networks either support the adaptation process or attract cities that are active" (Heikkinen et al., 2020).

2.1.4 Individuals within the organization

"Promoters of diffusion" can take on various shapes. Apart from things such as financial incentives and procedures, they can also include individuals (Kuhlmann et al., 2019). The influence of one individual (a lobbyist, civil servant, activist, or resident) can be quite large (Rogers, 1983). They can contribute to agenda setting, for example. In their 2017 study, De Ruiter and Schalk (2017) found that members of parliament often use policies from different countries in debates to highlight these initiatives to the public and pitch them as potential solutions for their own country. However, the specific role and influence of these individuals are difficult to control and include in a larger-scale study (Ryan, 2015). The same goes for individual policymakers/civil servants and their personal interests or intrinsic motivation for engaging with a certain topic. The inherent characteristics of policymakers can also play a role in the likeliness of policy adoption. Rashidi and Patt (2018) found that cities that employ more women are more likely to adopt climate policy.

2.1.5 Politics

Political context in terms of alignment with the policy of the governing entity, as well as the strength of lobby groups on the topic, has also been identified as an important motivator for policy adoption (Matisoff & Edwards, 2014). In some studies, political alignment with the national-level government, even when the relevant policy is not controversial, is significant when it comes to policy adoption (DellaVigna & Kim, 2022; Jans et al., 2016). A government unit's level of autonomy, in this case, the municipal level, is also important to be able to observe policy adoption (Zhang & Zhu, 2019). The type of (electoral) system can also have an effect on (the speed of) the adoption of certain policies (Orellana, 2010; Verschueren, 2022).

2.1.6 Other factors

There are many other possible internal factors contributing to a municipality adopting a policy; these include, but are not limited to, the sociodemographic and geographic features of a city. However, the ones mentioned before are the factors most commonly described in the literature. The variable for problem pressure, which is based on an index developed by Leidelmeijer and Mandemakers (2022), includes a large number of factors. These factors include physical environment, housing, amenities, social cohesion, and safety. This allows for the indirect inclusion of these variables.

2.2 External factors for policy diffusion

Policy diffusion theory is a widely studied field and views policymaking entities as interdependent. It focuses on the process of why entities, most often nation-states, come to adopt policies as a result of being influenced by other entities (Kuhlmann et al., 2019). This influence can be either horizontal (from nation to nation or from municipality to municipality, for example) or vertical (top-down from national-level government to subnational level unit of government, for example) (Shipan & Volden, 2008). De Ruiter and Schalk (2017) wrote that on the international level, policy interdependence is higher in some policy areas than in others. An example is environmental policies. This is because in environmental or climate policies, the effects of adopting or not adopting a policy cross geographic borders. An example is air pollution policy and the consequences of acting/not acting.

Policy diffusion requires interaction between units of governance. There are various models explaining the manner in which these interactions between governance units take place. Berry and Berry (2007) summarized them into the following five models based on the existing literature: the first one refers to the "National Interaction Model." This model focuses on the interactions between individuals (e.g., members of parliament and government officials) resulting in policy diffusion. As more individuals learn about something, the process of policy diffusion increases over time. The second summary model is the "Regional Diffusion Model," which focuses on geographic proximity as a prime mechanism of interaction between governance units and policy diffusion. The third model is the "Leader-Laggard Model," which is based on the assumption that certain states are early adopters or pioneers and other states follow suit as they consider them good examples (Walker, 1969). The fourth model is the "Isomorphism Model," a model that focuses on similarities between states and the likeliness of states to emulate other states that share similarities with them. The fifth and last model is the "Vertical Influence Model," which focuses on policy diffusion because of one leading agency that demands a certain policy to be adopted: vertical policy diffusion.

In all these models, interaction needs to take place between the governance units at stake. Practical examples of mechanisms of "interaction" are countries' official interactions during international forums and conferences with either countries that have already adopted a certain policy, or with countries that are similar in structure. Country structure can be similar in the sense of, for example, the type of diplomatic relations they have and with which countries, which points to "interaction similarity" (Kammerer & Namhata, 2018). An example of "interaction similarity" is found by De Ruiter and Schalk (2017), who, in their study on the effect of member of parliaments' references to other nations' policy adoptions, found that members of parliament in the Netherlands especially refer to countries of large size that are seen as influential and good examples (Haupt et al., 2020) or that are also part of the European Union. Lee and van de Meene (2012) found that cities especially learn from cities with shared languages or geographies. Matisoff and Edwards (2014) used Walker (1969) state groupings in their research, which are largely based on similarities in the likeliness of policy adoption due to political preferences and attitudes. They found that these states and their similar acting have stayed relatively stable over decades and that their shared political culture is the most accurate predictor of policy adoption as opposed to proximity.

For that reason, there is an increasing focus on studying non-Western countries in policy diffusion theory as these countries often are structurally different and interact with other similar nations, as well as in different ways (Zhang & Zhu, 2019). However, not only is it important what countries look at other countries for policy, but also that these "pioneer countries" have a high capacity and sufficient resources in policymaking to allow them to be a pioneer in the field (Heikkinen et al., 2020; Jänicke, 2005). Cities with more resources available are also more likely to join a network (Heikkinen et al., 2020). Networks are an important way in which interaction leading to policy diffusion takes place. The most common external factors for policy adoption and diffusion as discussed in the academic literature are as follows.

2.2.1   Municipal networks

The effect of networks on the adoption of policy is not well-researched (Rashidi & Patt, 2018). However, there is research that has found a positive effect of transnational networks on the likeliness of adopting (climate) policy (Lee & Koski, 2014), also in the case of trans-municipal networks specifically (Häußler & Haupt, 2021; Nguyen et al., 2020). Networks can play both a practical role, by promoting the exchange of information or resources, and a political role. The political role of a network expresses itself through the creation of political support as well as highlighting the city's engagement with the topic (Heikkinen, 2022). Some international trans-municipal networks that have been found to play a role in policy adoption are the C40 network and the ICLEI (Rashidi & Patt, 2018). Other networks that have been included in policy diffusion and adoption research are the United Nations International Strategy for Disaster Reduction, EURO-CITIES, Citynet, Energy Cities, Climate Alliance, and URBACT (Haupt & Coppola, 2019).

Transnational municipal networks, in the case of the 100 Resilient Cities Network, for example, can be innovative in terms of the tools they use to bring about change (Papin, 2019). It is, however, still dependent on internal factors whether or not a municipality is able to take full advantage of what such a network can provide (Bellinson & Chu, 2019). Small- and medium-sized cities are less likely to participate in such networks because of a lack of resources but could benefit a lot from "pooled resources" in networks specifically for small- and medium-sized cities, as expectations are different in such networks (Häußler & Haupt, 2021). In addition to that, participation in a network where knowledge exchange takes place does not automatically result in learning (Haupt et al., 2020). As cities often participate in multiple networks, both internationally and nationally, it is difficult to control the effects of a particular network. In addition to that, learning and exchange are not limited to formal networks only but can take place in many ways, even without direct interaction. Scholars still advocate for more research on the effect of city-to-city learning and trans-municipal networks (Fünfgeld, 2014; Haupt et al., 2020; Haupt & Coppola, 2019).

2.2.2 Geographic proximity and similarities

The role of geographic proximity, as Berry and Berry (2007) identify as one of the main models that has historically been used in explaining policy diffusion, has been debated in more recent scholarly research. Its influence on policy adoption has decreased over time (DellaVigna & Kim, 2022). A possible reason is that the cost of transportation and communication has been significantly decreasing over the past few decades, since when policy diffusion research first came to be (Matisoff & Edwards, 2014). Scholars argue for better research on the different types of relationships between policymakers as opposed to geographic proximity (Carley & Nicholson-Crotty, 2018), and many studies find no strong results in terms of geographic proximity (Zhang & Zhu, 2019; Zhou et al., 2019), but economic proximity, or similarity, does seem to have an effect (Zhou et al., 2019). This is similar to the "interaction similarity," as Kammerer and Namhata (2018) posit. Cities look to cities that face similar challenges (Haupt et al., 2020). Documenting all the different types of relationships that exist among geographic units is very difficult, but studies with a small sample size have been carried out (Carley & Nicholson-Crotty, 2018).

2.3 Underlying motives and research gaps

While interactions are the manner in which policy spreads among geographic units, the interactions must be fueled by underlying motives to result in policy diffusion. In a comprehensive literature review, Danaeefard and Mahdizadeh (2022) found that scholars have identified four "causal mechanisms" that may lead to policy diffusion between countries: competition, emulation, learning, and coercion. Obinger et al. (2013) described them as follows: the first mechanism, *learning*, describes countries adopting policies that have already been successful in other countries. Specifically in countries that faced similar problems in the past or are currently facing similar issues or problems. The policy these countries have already developed and adopted provides solutions to the problem, and it makes sense for the other countries to make use of this solution and copy it. *Emulation,* sometimes referred to as *imitation,* describes countries that adopt policies as they have ambitions to be part of an "international norm-based community" in which the "symbolic and socially constructed meaning" of policy adoption is more important than the content of the policy itself (Yi & Liu, 2022). This relates to the "interaction similarity" where countries mostly interact with countries similar to them (De Ruiter & Schalk, 2017; Kammerer & Namhata, 2018), or the Isomorphism model as identified by Berry and Berry (2007). *Competition* aims at the wants and needs of governments to compete with their counterparts and thus feel the incentive to adopt similar policies (Obinger et al., 2013). Depending on the type and sector of the policy, this competition can be in terms of economic development or ensuring economic competitiveness, for example (Saikawa, 2013; Zhang & Zhu, 2019). *Coercion* describes the imperative measures that countries are required to implement, such as coming from the European Union or the United Nations (Obinger et al., 2013), pointing at the Vertical Diffusion Model, as summarized by Berry and Berry (2007).

In their research, policy diffusion scholars have focused on different combinations of the mechanisms described above (Abel, 2021; Bergero et al., 2021). Zhang and Zhu (2019) found a combination of learning and emulation as important drivers for policy diffusion. Gilardi and Wasserfallen (2019) posited that this framework makes certain assumptions, including that decision-makers make fact-based decisions and that decision-makers are rational actors (Kousky & Schneider, 2003; Yi & Liu, 2022).

The mechanisms put forward by policy diffusion theory (learning, emulation, competition, and coercion) make up the dominant framework that is utilized when describing the interdependence of entities adopting policy from other entities. However, as described above, there are both internal and external factors driving policy adoption in municipalities (Gordon, 2013; Kammerer & Namhata, 2018; Matisoff & Edwards, 2014). These mechanisms have also been proven to be conditional on certain internal factors, as described above (Mallinson, 2021). Scholars sometimes focus their research on which are more important—internal or external factors, but there is no consensus among them (Fuentes & Pipkin, 2022).

Policy adoption and diffusion have often been studied on a global level, comparing different countries and their interactions (Kammerer & Namhata, 2018). Scholars have also looked at this on a state- or province-level in large countries such as the United States (Carley & Nicholson-Crotty, 2018; Matisoff & Edwards, 2014; Pereira, 2022) and China (Zhang & Zhu, 2019). However, studies are limited on countries that are of smaller size (such as the Netherlands) and on smaller subnational scales (such as the municipality-level).

Policy diffusion research regarding resilience policy specifically has not yet been carried out, but certain parallels between resilience policy and climate policy can be drawn. Climate policy makes up an important part of resilience policy and often overlaps with resilience policy: resilience policy also targets climate issues and seeks to reach climate resilience. In addition to that, policies focusing on resilience are, similar to climate policies, generally long-term policies (Wang, 2021). They are often without immediate effects, and cities do not adopt them expecting short-term results. These policies can therefore also not be used for short-term political gain. Scholars such as Salvia et al. (2021) call for more research on motivators for climate policy adoption, such as the stresses a city faces. Research addressing the motivators of resilience policy adoption can also be relevant to climate policy adoption research. However, while resilience policy indirectly includes climate policy, this chapter focuses on resilience policy only.

In addition to that, current resilient research is largely case-study focused. There have been many case studies on individual or small sets of (European) cities that indicate that the adoption of a resilience framework has had a positive effect on responding to climate threats (Lu & Stead, 2013), but a larger-scale assessment of resilience policy and its adoption in Europe is not yet available. The exceptions focus on the theoretical approach to resilience or its implementation rather than policy innovation or adoption (Shamsuddin, 2020).

According to the OECD, cities often set climate and resilience goals that are more ambitious than those of national governments (OECD, n.d.-a). This is often by necessity: urban residents already notice the effects of climate change. But this is also by nature: cities can tailor-make specific policies, whereas for national governments this is often more difficult (Short, 2015). It is therefore relevant to study the adoption of resilience policy at the city-level.

Many cities face many difficulties regarding the adoption and implementation of resilience policy (Coaffee et al., 2018; Huck et al., 2020), which can be pointed to the lack of institutional incentives or the need for intersectoral collaboration, among other things (Coaffee et al., 2018; Huck et al., 2020; Lee, 2013). However, there are many international programs targeting to assist local governments and municipalities in adopting resilience policies.

2.4 Case study: The Netherlands

2.4.1 Theoretical framework and research questions

Research on municipal resilience policy in the Netherlands has not yet been carried out.

As described in the literature review, there are numerous factors driving policy adoption. By researching what have been the drivers for cities that have already integrated resilience into their policy, more insights can be found into how more cities can be incentivized to adopt resilience policy and how designated resilience programs can better provide assistance to cities. Resilience policy in this chapter is defined as a strategy that acknowledges becoming resilient as something beneficial and seeks for the unit of government to become resilient.

Although certain Dutch cities, especially Rotterdam and Amsterdam (Huck et al., 2020, 2021; Sharma, 2022), have been studied as part of their participation in transnational resilient city networks or their resilience policy on a case-study basis, no larger-scale assessments of the adoption of resilience policy in the Netherlands have been performed. The literature review discussed the drivers for policy adoption according to the existing academic literature on policy diffusion. Based on that literature, policy diffusion is defined as the adoption of a policy by one unit as a result of influence by another unit (Gilardi & Wasserfallen, 2019). Whether a factor is categorized as an internal factor or an external factor is described in Table 1.

As described in the literature review, the academic literature finds no single factor that motivates policy adoption or not, but it is rather a combination of internal and external factors. Many studies combine both the external and internal drivers of policy adoption in their models (Matisoff, 2008; Rashidi & Patt, 2018; Zhou et al., 2019). Based on these existing studies, we therefore assume that both external and internal drivers play a role in the adoption of resilience into policy for cities in the Netherlands (Fig. 1).

However, this research project seeks to find out what is a more important driver of policy adoption, focusing on the resilience policy adoption by the 50 largest municipalities in the Netherlands, internal or external factors.

Four hypotheses were formed in order to find the answer to the final research question. These hypotheses were formed based on a broad range of existing academic literature on the adoption and diffusion of policy. This research has been found in regard to a wide variety of policies in different policy areas; thus, the assumption was made that similar relationships would be found in regard to resilience policy.

Based on the academic literature focusing on the effect of internal factors on policy adoption (Gordon, 2013; Kammerer & Namhata, 2018; Matisoff & Edwards, 2014), specifically the role of problem pressure (Matisoff & Edwards, 2014), and the assumption that context is important (Schoenefeld et al., 2022) and that policymakers make fact-based decisions (Gilardi & Wasserfallen, 2019), the first hypothesis is as follows:

Hypothesis 1 Cities that are at higher risk for the stresses resilience policy aims to address are more likely to have adopted resilience policy/integrated resilience as a concept in policy.

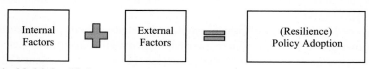

FIG. 1 Model simplified.

Based on the academic literature focusing on the effect of internal factors on policy adoption (Gordon, 2013; Kammerer & Namhata, 2018; Matisoff & Edwards, 2014), especially the research on the influence of available (financial and human) resources on policy adoption within municipalities (Aguiar et al., 2018; Araos et al., 2016; Jans et al., 2016; Matisoff & Edwards, 2014; Otto et al., 2021; Ryan, 2015), the second hypothesis is as follows:

Hypothesis 2 Cities with more resources are more likely to adopt resilience policy/integrated resilience as a concept in policy.

Based on research regarding the impact of globalization on policy adoption (Lee, 2013; Lee & Koski, 2014) and emulation as a driver (imitation and desire to be part of a "norm-based community") for policy adoption (Yi & Liu, 2022; Zhang & Zhu, 2019), the third hypothesis is as follows:

Hypothesis 3 Cities that have a higher level of globalization are more likely to adopt resilience policy/integrated resilience as a concept in policy.

Utilizing

1. the academic literature focusing on policy diffusion and the effect of external factors on policy adoption (Berry & Berry, 2007; Kuhlmann et al., 2019; Shipan & Volden, 2008; Walker, 1969);
2. the academic literature on the role of networks and the evidence that it can incentivize policy adoption (Häußler & Haupt, 2021; Lee & Koski, 2014; Nguyen et al., 2020);
3. the research in which participation in policy networks in the Netherlands specifically has proven to play an important role in early adoption of a policy (Jans et al., 2016), the fourth hypothesis is as follows:

Hypothesis 4 Cities involved in trans-municipal networks are more likely to adopt resilience policy/integrated resilience as a concept in policy.

3 Methodology

This study focuses on the 50 largest municipalities in the Netherlands. These municipalities were chosen because there is a lot of variety between them in terms of the variables (resources, level of globalization, etc.) that are included in this study, as well as geographic spread throughout the Netherlands. Fifty municipalities are approximately 17% of the total number of municipalities, and the population numbers of the 50 largest municipalities in the Netherlands add up to a large percentage of the total population of the Netherlands. Focusing on this particular population means that the findings are not generalizable to all municipalities in the

The methodology of this project employed mixed methods and consists of the following two stages: qualitative conceptual content analysis and quantitative statistical analysis with the following steps: (1) Policy Document Collection, (2) Policy Document Analysis, (3) Municipality Classification, and

(4) Statistical Analysis. Prior to performing content analysis, data were collected manually by collecting and selecting relevant policy documents from each of the 50 municipalities (see other chapters) in the following three steps: (1) collection of thematic policy documents on resilience through the municipal website, (2) web search on municipality and resilience, and (3) collection of strategic policy documents part of the municipal policy cycle through the appropriate municipal channels. Following the collection of the documents, the collected policy documents were analyzed qualitatively (Can this be defined as resilience policy according to the definition of this study? Which of the four drivers of resilience (economy, government, society, and environment) does this document address? Does this document integrate the concept of resilience into strategic policy?). The results were then used to classify the municipalities accordingly. After collecting data on the 50 municipalities and classifying them, the data were used to perform statistical analysis to find out what the most important drivers of policy adoption are: internal or external factors.

3.1 Empirical strategy

3.1.1 Data measurement

The independent variable "Stress Level" refers to data collected through the liveability index created by Leidelmeijer and Mandemakers (2022). The liveability index model has been used over several years to measure liveability and its development over the years on the neighborhood level in Dutch cities. The model uses a total of 94 different variables to measure livability within five different dimensions: physical environment, housing, amenities, social cohesion, and safety. It not only includes variables such as proximity to nature and proximity to water but also environmental risks such as earthquake proneness, heat stress, flood risk, and air quality. It is the third model of its kind and has been periodically updated to include the most relevant and up-to-date variables and data. The date of measurement of the data in this report is January 1, 2020 (Table 2).

These data are utilized for assessing the problem pressure of a municipality since the broad and intersectoral nature of the liveability index and the variables it includes align with the interdisciplinary character of resilience: "the ability to prepare for and handle future shocks and stresses on the economy, government,

TABLE 2 Independent variable: stresses.

Independent variable	Data source
Stress	Leidelmeijer and Mandemakers (2022)
Stress quantity	Leidelmeijer and Mandemakers (2022)

society, and environment" (Coaffee et al., 2018; OECD, n.d.-b; Sharma et al., 2023). In addition to that, data availability also played a role in choosing this dataset since data on the individual characteristics part of the liveability index (Leidelmeijer & Mandemakers, 2022) on the municipal level are not available for public use.

The data used for the variable "Stress Level" specifically refers to whether or not a city has areas that received a low score on the liveability meter over time and are classified in the report as being under structural stress. It is an indicator, or binary categorical, variable with a nominal level of measurement. The variable "Stress Quantity" refers to the percentage of total housing units in a city that is structurally under stress and is a continuous variable with a ratio level of measurement.

Resources can refer to different types of resources. This study will look at human and financial resources. The variable "Population Size" refers to the population size of the municipality in real numbers. The data come from the Netherlands Central Bureau for Statistics and are taken from the most recent year available, 2020. Population size is a proxy for human resources, based on the assumption that larger cities employ more people as they have more resources available and more people they need to provide for. Population size is utilized as a variable because of the absence of data on the number of municipal employees. It is a continuous variable with a ratio level of measurement (Table 3).

The variable "Municipal Budget" refers to the financial budget of the respective municipality according to the most recent available data, 2021. The data source is FinDo, a government source. This accurately reflects the number of monetary resources available to a municipality. It is a continuous variable with a ratio level of measurement (see Table 9).

In the absence of a variable measuring the level of globalization of a city, as a proxy for globalization, the following three variables are utilized: the first variable "English" refers to whether or not a municipality has a website in English. Despite the fact that the Netherlands counts many non-Dutch speakers, it is not standard in the Netherlands that a municipal website is available in English. If an English website is available, this points to the municipality wanting to be part of the international community by being accessible to the international

TABLE 3 Independent variable: resources.

Data	Data source
Population size	CBS (2022)
Municipal budget	FinDo (2022)

TABLE 4 Independent variable: globalization.

Data	Data source
English	Municipality websites
Global goals	VNG (n.d.-b)
University	University websites

community. This variable is an indicator, or binary categorical, variable with a nominal level of measurement (Table 4).

The second variable "Global Goals" refers to whether or not a municipality has chosen to be a "Global Goals municipality." A municipality in the Netherlands can voluntarily choose to align with this UN framework and profile itself as such. This also points to a municipality wanting to be part of the international community. "Global Goals" is an indicator, or binary categorical, variable with a nominal level of measurement.

The third variable, "University," points to whether or not a city has a university within the city's borders. Universities operate in the international realm and often bring about international collaboration and visitors to a city. "University" is an indicator, or binary categorical, variable with a nominal level of measurement.

According to the Association of Netherlands Municipalities (VNG), of which all Dutch municipalities are part, there are four main networks: the G4 network with membership of the four largest cities in the Netherlands, the G40 network with membership of municipalities with more than 100,000 inhabitants, the M50 with member cities having between 30,000 and 80,000 residents, and the P10 network for large rural municipalities (VNG, n.d.-a). A recent development among municipalities is the Metropolitan Regions. These are networks of cities geographically located around major cities. There are three Metropolitan Regions in the Netherlands: Amsterdam (30 member cities), Rotterdam-The Hague (21 member cities), and Eindhoven (21 member cities). The specific topics and scope of what the Metropolitan Regions work on differ per Metropolitan Region, but the main purpose of the existence of the Metropolitan Regions is to tackle strategic issues that cross municipal borders such as infrastructure and public transport, but also economic growth and innovating the economic structure (MRA, n.d.; MRDH, n.d.; MRE, n.d.). Working together as one region also makes it easier to compete with other regions or metropoles globally (MRDH, n.d.) and provides more capacity to work on metropolitan challenges (MRE, n.d.). This network will also be included in the study (Table 5).

The variable networks refer to whether or not the respective city is part of one of the three networks included in this study: the G40, the M50, and the

TABLE 5 Independent variable: networks.

Independent variable	Data source
Network membership: G40	G40 (n.d.)
Network membership: M50	M50 (n.d.)
Network membership: metropolitan region	MRA (n.d.), MRDH (n.d.), and MRE (n.d.)
Network membership: combined	G40 (n.d.), M50 (n.d.), MRA (n.d.), MRDH (n.d.), and MRE (n.d.)

Metropolitan Regions, as well as a combined variable if a city is part of any of the three networks. All of these variables are indicator/binary categorical variables with a nominal level of measurement.

3.1.1.1 Control variables

For external factors, the control variable is province, indicating which of the 12 provinces the city is located in. Provinces can be an indicator of structural similarities between states, following Walker (1969) model on state groupings. Province is a nominal categorical variable (see Table 6).

3.1.1.2 Dependent variable

The dependent variable is resilience adoption and indicates the degree of resilience adoption. It is an ordinal categorical variable: each of the municipalities is assigned a value according to their classification from 1 to 4, as described in Table 4. This classification was intentionally created based on the early stages of the conceptual content analysis. These classifications were necessary to accurately depict the level of adoption of the concept of resilience into policy, because of the large differences in the extent to which they did so: all municipalities did adopt the concept of resilience into their policy, but only a few did so in an overarching policy plan. For that reason, the dependent variable will be measured as an ordinal categorical variable in order to allow for those differences to be part of the model.

In Table 6, we find an overview of all the variables included in the model.

3.1.2 Regression model

As the data were collected within the same time frame, the first month of 2023, the chosen method of statistical analysis is cross-sectional regression. Cross-sectional logistic regressions are commonly performed in regard to policy diffusion and policy adoption research (Henley & Lee, 2023). Event history analysis (EHA) is also a common method for policy diffusion research (see, for

TABLE 6 Variables overview.

Variables	Data type	Label	Level of measurement
Stress (independent)	Indicator/binary categorical	Areas under stress yes (1)/no (0)	Nominal
Stress quantity	Indicator/categorical	Percentage of total housing units under structural stress divided in categories	Ratio
Population size (independent)	Continuous	Population size in real numbers	Ratio
Municipal budget (independent)	Continuous	Municipal budget in real numbers	Ratio
English website (independent)	Indicator/binary categorical	English website yes (1)/no (0)	Nominal
Global goals (independent)	Indicator/binary categorical	Global goals municipality yes (1)/no (0)	Nominal
University (independent)	Indicator/binary categorical	University in city yes (1)/no (0)	Nominal
Network membership: G40 (independent)	Indicator/binary categorical	Member of G40 yes (1)/no (0)	Nominal
Network membership: M50 (independent)	Indicator/binary categorical	Member of M50 yes (1)/no (0)	Nominal
Network membership: metropolitan region (independent)	Indicator/binary categorical	Member of metropolitan region yes (1)/no (0)	Nominal
Network membership: combined (G40, M50, metropolitan region)	Indicator/binary categorical	Member of any network yes (1)/no (0)	Nominal
Province (independent)	Nominal categorical	Which of the 12 provinces in the Netherlands	Nominal
Resilience class (dependent)	Ordinal categorical	Resilience classification	Ordinal

example, Berry & Berry, 2007; Matisoff & Edwards, 2014; Zhang & Zhu, 2019; Zhou et al., 2019). However, in order to employ EHA, time-series data for the adoption of a certain policy need to be available. As resilience policy is relatively new, this project limits its focus to cross-sectional data.

Equation 1. Model

$$\text{Resilience policy adoption} = \beta_0 + \beta_1 \text{ internal factors} + \beta_2 \text{ external factors} + Ui$$

$$=$$

$$\text{Resilience policy adoption} = \beta_1 \text{ resources} + \beta_2 \text{ problem pressure} + \beta_3 \text{ globalization} + \beta_4 \text{ network participation} + Ui$$

$$(1)$$

β_1 are the continuous variables "Population Size," as a proxy for municipal employees, and "Municipal Budget." β_2 describes the variables "Stress" and "Stress Quantity." β_3 refers to the proxy variables for globalization: "English," "Global Goals," and "University." β_4 includes the variables "Metropolitan Region," "G40," "M50," and "Network Membership Combined." This model makes the assumption that internal and external factors are equally important and is therefore a linear model.

The logistic regression analysis is performed using STATA software version 17. First, an ordered logistic regression has been performed. This regression type uses maximum likelihood estimation (Braverman et al., 2018) to calculate the probability of resilience policy adoption/integration of the concept of resilience into policy based on the included independent variables. This type of regression was chosen because the dependent variable, degree of resilience according to classification, is an ordinal categorical variable. The distance between the different groups, the classification from 1 to 4, is not equal. Due to the data characteristics, ordered logistic regression is the most relevant method as the model does not "assume equal intervals between scoring categories" (Stewart et al., 2019). Ordered logistic regression is a method that has been employed in research on policy diffusion (Lee, 2013; Lee & Koski, 2014) and policy evaluation (Braverman et al., 2018).

4 Results

4.1 Summary statistics

Tables 7–10 provide the summary statistics for the variables included in the study.

4.2 Summary statistics

In Table 11, we find that the ordered logistic regression shows a significant relationship for the variables "English," "Stress Category," and "G40." These

TABLE 7 Summary statistics independent categorical variables.

Variable	Frequency	Percentage	Cumulative
Areas under structural stress			
No	19	38	38
Yes	31	62	100
Stress category			
0%–9%	20	40	40
10%–19%	12	24	64
20%–29%	12	24	88
30%–39%	5	10	98
40%–49%	1	2	100
Website in English			
No	17	34	34
Yes	33	66	100
Global goals municipality			
No	14	28	28
Yes	36	72	100
University			
No	36	72	72
Yes	14	28	100
Metropolitan region			
No	31	62	62
Yes	19	38	100
G40			
No	9	18	18
Yes	41	82	100
M50			
No	47	94	94
Yes	3	6	100

TABLE 8 Summary statistics independent continuous variables.

Variable	Observations	Mean	Standard deviation	Minimum	Maximum
Population size	50	164502.34	152504.11	76,653	903,929
Municipal budget	50	862993.82	1247242.3	238,325	7,999,801

TABLE 9 Summary statistics control variable.

Variable	Frequency	Percentage	Cumulative
Province			
Drenthe	1	2	2
Flevoland	2	4	6
Friesland	2	4	10
Gelderland	4	8	18
Groningen	1	2	20
Limburg	4	8	28
Noord-Brabant	9	18	46
Noord-Holland	9	18	64
Overijssel	4	8	72
Utrecht	2	4	76
Zuid-Holland	12	24	100

TABLE 10 Summary statistics dependent variable.

Variable	Frequency	Percentage	Cumulative
Resilience classification			
Classification 1	0	0	0
Classification 2	27	54	54
Classification 3	20	40	94
Classification 4	3	6	100

TABLE 11 Regression results I.

		Ordered logistic regression				
Resilience class	Coef.	St. err.	t-value	P-value	[95% conf Interval]	Sig
Part of G40	0					
Yes	−2.611	1.319	−1.98	.048	−5.196 to −.027	**
Part of M50	0					
Yes	−17.362	1334.505	−0.01	.99	−2632.944 to 2598.221	
Part of metro	0					
Yes	−.193	.8	−0.24	.809	−1.762 to 1.375	
Province	.178	.131	1.36	.173	−.078 to .435	
Global goals	0					
Yes	−.325	.964	−0.34	.736	−2.214 to 1.565	
English website	0					
Yes	2.395	1.081	2.21	.027	.275 to 4.514	**
University	0					
Yes	.115	.932	0.12	.902	−1.711 to 1.941	
Stress category	.084	.035	2.40	.016	.015 to .153	**
Population size	0	0	1.18	.239	0	
Municipal budget	0	0	−0.84	.402	0	

Continued

TABLE 11 Regression results I—cont'd

			Ordered logistic regression			
Resilience class	Coef.	St. err.	t-value	P-value	[95% conf Interval]	Sig
cut1	2.296	1.536	.b	.b	−.715 to 5.307	
cut2	6.464	1.97	.b	.b	2.602 to 10.326	
Mean dependent var.		1.560	SD dependent var.			0.611
Pseudo r-squared		0.293	Number of obs.			50
Chi-square		25.705	Prob>chi2			0.004***
Akaike crit. (AIC)		85.956	Bayesian crit. (BIC)			108.900

*** $P < .01$, ** $P < .05$, * $P < .1$.

specific results are significant because we find that the P-value for "G40" is .048, the P-value for "English" is .027, and the P-value for "Stress Category" is .016. P-values smaller than 0.05 indicate statistically significant results and thus reject the null hypothesis. The P-values for the other variables are higher than .05 and thus do not indicate significant results.

The relationships between resilience and the tested variables are positive for "English" and "Stress Category" and negative for "G40": The coefficient of the variable "G40" is -2.611239. This means that being part of the G40 gives a -2.611239 lower log odd of being in a higher classification and thus a higher probability of having adopted resilience policy or integrated the concept into policy. The coefficient of the variable "English" is 2.394523. This means that having a website in English gives a 2.394523 higher log odd of being in a higher classification and thus a higher probability of having adopted resilience policy or integrated the concept into policy. The coefficient of the variable "Stress Category" is 0.0839648. This means that for each increase in stress category, there is a 0.0839648 higher log odd of being in a higher resilience classification and thus a higher probability of having adopted resilience policy or integrated the concept into policy.

At the bottom of the output, we find that our total number of observations is 50. The likelihood ratio chi-square is 25.71 with a P-value of .0042. The value of the pseudo-R-squared is 0.2932. Overall, this means that the model is statistically significant. A likelihood ratio test has also been performed to test the validity of the model and results. However, when testing for collinearity, it is found that the proxy variables for globalization and the variables for networks show correlation. When omitting these variables from the model, the model significance remains similar, but the "Stress Category" is the only variable that consistently shows significant results.

In Table 12, we find that when omitting the variables for network in the regression, the coefficient of the variable "Stress Category" is 1.077828. This means that for each increase in stress category, there is a 1.077828 higher log odd of being in a higher resilience classification and thus a higher probability of having adopted resilience policy or integrated the concept into policy.

At the bottom of the output, we find that our total number of observations is 50. The likelihood ratio chi-square is 18.62 with a P-value of .0095. The value of the pseudo-R-squared is 0.2124. Overall, this means that the model is statistically significant. A likelihood ratio test has also been performed to test the validity of the model and results.

In Table 13, we find that when omitting the variables for globalization in the regression, the coefficient of the variable "Stress Category" is 1.069047. This means that for each increase in stress category, there is a 1.069047 higher log odd of being in a higher resilience classification and thus a higher probability of having adopted resilience policy or integrated the concept into policy.

TABLE 12 Regression results II.

						Ordered logistic regression	
Resilience class	Coef.	St. err.	t-value	P-value	[95% Conf Interval]		Sig
Province	1.171	.144	1.29	.198	.921 to 1.489		
Global goals							
Yes	.594	.453	−0.68	.494	.133 to 2.646		
English website							
Yes	3.456	2.833	1.51	.13	.693 to 17.228		
University							
Yes	1.271	1.079	0.28	.778	.241 to 6.707		
Stress category	1.078	.032	2.50	.012	1.016 to 1.143		**
Population size	0	0	1.32	.187			
Municipal budget	0	0	−0.97	.331			
cut1	3.629	1.458	.b	.b	.772 to 6.486		
cut2	7.477	1.894	.b	.b	3.766 to 11.188		

Mean dependent var	1.560	SD dependent var	0.611
Pseudo r-squared	0.212	Number of obs	50
Chi-square	18.619	Prob>chi2	0.009***
Akaike crit. (AIC)	87.042	Bayesian crit. (BIC)	104.251

*** P < .01, ** P < .05, * P < .1.

TABLE 13 Regression results III.

Resilience class	Ordered logistic regression					
	Coef.	St. Err.	t-value	P-value	[95% Conf Interval]	Sig
Province	1.091	.123	0.77	.441	.875 to 1.359	
Part of G40						
Yes	.339	.32	−1.15	.252	.053 to 2.16	
Part of M50						
Yes	0	0	−0.01	.991	0	
Part of metro						
Yes	.976	.701	−0.03	.972	.238 to 3.991	
Stress category	1.069	.034	2.12	.034	1.005 to 1.137	**
Population size		0	1.79	.073		*
Municipal budget		0	−1.27	.203		
cut1	1.837	1.354	.b	.b	−.817 to 4.491	
cut2	5.691	1.741	.b	.b	2.279 to 9.104	
Mean dependent var	**1.560**	**SD dependent var**				**0.611**
Pseudo r-squared	0.218	Number of obs				50
Chi-square	19.101	Prob>chi2				0.008***
Akaike crit. (AIC)	86.560	Bayesian crit. (BIC)				103.768

*** $P < .01$, ** $P < .05$, * $P < .1$.

At the bottom of the output, we find that our total number of observations is 50. The likelihood ratio chi-square is 19.10 with a P-value of .0079. The value of the pseudo-R-squared is 0.2179. Overall, this means that the model is statistically significant. A likelihood ratio test has also been performed to test the validity of the model and results.

5 Discussion

5.1 Discussion of results

5.1.1 Summary results

The data suggest that there is a positive correlation between the level of structural stress a city is in, thus the problem pressure, and the integration of resilience into policy. The quantitative analysis supports Hypothesis 1 of the paper: Cities that are at higher risk for the stresses resilience policy aims to address are more likely to have adopted resilience policy.

No evidence was found in support of Hypothesis 2, that cities with more resources are more likely to have adopted resilience policy or have integrated the concept of resilience into policy. No significant results were found in support of Hypothesis 3, that cities that have a higher level of globalization are more likely to have adopted resilience policy or have integrated the concept of resilience into policy. No evidence was found in support of Hypothesis 4, that cities part of trans-municipal networks are more likely to have adopted resilience policy or have integrated the concept of resilience into policy. The control variable province was also not found to be significant. Although evidence in support of only one hypothesis was found, these data support the central argument that internal factors are a more important driver of policy adoption than external factors.

5.1.2 Interpretation of results

Although the literature includes individuals and politics as important internal factors in policy adoption, the variables included in this chapter's model do not include politics and individuals. This is for reasons of data availability, chosen methodology, and data collection methods. Generally, the role of individuals is difficult to include in a model (Ryan, 2015). A different methodology, employing case studies and interviews, would be necessary in order to include information on the influence of individuals within a municipality and their adoption of resilience policy. The time in which this project needed to be completed did not allow for a study of a relatively large number of municipalities conducting interviews with each of them.

The significant results in regard to problem pressure are in line with expectations, as they are aligned with the literature that cities with increased stress, at increased risk, are more likely to adopt policies to prepare for these risks or to mitigate the effects of the stress they experience (Matisoff & Edwards, 2014).

This can be because, as Matisoff and Edwards (2014) discuss, of the expected problem-solving ability of the policy. Another potential explanation could be that programs targeting cities to help them adopt resilience policy, as introduced earlier in this paper, are effectively targeting cities under increased stress. Regardless of the underlying reason, it is a positive development that the cities that are likely to most benefit from the concept of resilience and its integration into policy, given their increased risk and stress, have in fact done so.

The absence of results in regard to a city's level of globalization is not in line with the existing literature (Lee & Koski, 2014). This could be due to the proxy variables ("English," "Global Goals," and "University") chosen to measure this variable. However, the variables globalization and network participation did indicate correlation in models utilized. This is in line with the literature that describes that level of globalization plays a role in likeliness of network participation (Heikkinen et al., 2020; Lee, 2013).

A city's level of resources does not show significant results, which is in contract with the existing literature (Aguiar et al., 2018). This is not in line with the existing literature and with the hypothesis based on the literature. This is surprising, especially since the degree to which resources play a role in the likeliness of policy adoption depends on the policy (Ryan, 2015). As resilience is an interdisciplinary policy that requires municipalities to work cross-sectoral, it is expected that its adoption and implementation require a lot of resources.

A possible explanation for the absence of significant results in regard to resources is that the study included the 50 largest municipalities in the Netherlands. Although there is variance between the cities in regard to population size (proxy for human resources) and financial resources, there would be more variance if the study included smaller cities as well. Both because of the financial and human resources available for the process of policy adoption, but also because they are less likely to participate in networks (Fünfgeld, 2014; Haupt et al., 2020; Häußler & Haupt, 2021). In this sample, most cities were participating in the networks included in the study.

Although many scholars have highlighted the role of external factors in prior academic research, the results do not support these claims. No significant results were found in regard to networks, with the exception of a small negative correlation between participation in the G40 network and the integration of resilience into policy. This is an unexpected result and does not fit in with the existing academic literature and the policy diffusion literature that places a lot of importance on the role of networks as a driver for policy diffusion and adoption. Especially since within municipalities in the Netherlands specifically, the role of networks in policy adoption has previously been found to be significant (Jans et al., 2016). This study's findings do not support these claims. The inclusion of smaller cities in the research could show different results as most of the cities included in this study were part of a network.

A possible explanation is that, as mentioned in the study, there are numerous networks and collaborations that municipalities in the Netherlands take part in.

These are both within the Netherlands as well as transnational. Only three municipalities included in this study were not part of any networks, and several municipalities were part of two networks included in the study. In addition to that, the literature is unclear whether "networks either support the adaptation process or attract cities that are active" (Heikkinen et al., 2020).

In addition to that, this study is of cross-sectional nature. Politics shift over time on both the national level and within a municipality. The exact moment of adoption of the resilience policy by a municipality was not measured. Instead, the measurement of resilience policy was done on the basis of whether having adopted resilience policy or not. In addition to that, this study was carried out as municipal elections had just taken place. It was therefore not possible to include a variable on politics as it would have a high chance of being inaccurate.

5.2 Limitations and future research recommendations

First, although the data collection and analysis have been carried out in a careful manner, there is a chance human error will occur. In addition to this, the sample size of this study is small since it included only 50 municipalities in the Netherlands. However, due to the time-intensive nature of the data collection, this was the possible scope of the study. Although the municipalities are diverse in terms of the included variables in these studies, a study with inclusion of more municipalities may gather different results. This choice of population makes it that the findings are less generalizable. This is the major limitation of this study. This means that future studies could include all municipalities in the Netherlands and their integration of resilience into policy.

In addition to that, a correlation between participation in networks and a city's level of globalization was found. This was not the scope of this research, but future research could look into this correlation and its relation to policy adoption. In terms of measurement, the variables globalization and networks were not comprehensive. Since the variable globalization is not directly observable, proxy variables (English, Global Goals, and University) were utilized to measure globalization. As mentioned before, due to data availability, the variable networks only included the most important networks in the Netherlands and not all the networks the municipalities take part in.

Another limitation is that the aforementioned factors in explaining resilience policy adoption by municipalities, and those included in the study, are not all-encompassing. As mentioned previously, individuals can play an important role in policy- and decision-making (Blatter et al., 2021; Pereira, 2022; Yi & Liu, 2022). Variables for the role of politics and individuals within the organization are drivers that were also discussed in the existing literature but were not included in the study due to difficulties of data collection and variable inclusion. This is because of the cross-sectional nature of the study: since political alignment shifts over time within a municipality, it was not possible to include this as a variable. Since this study did not include interviews as a method, it was not possible to gather data on the role of specific individuals or networks in the resilience-adoption process. Future studies can choose to employ a qualitative analysis approach with interviews to find out more about underlying motives and specific interactions between units of governance through, for example, networks.

As mentioned previously, this research project only uses cross-sectional data as resilience policy is relatively new and the adoption process is ongoing. Because resilience is a relatively new concept and municipalities generally adopt policies at a different time, it is possible that in the future, more policies will have a resilience plan or have integrated the concept into policy. This is in line with research indicating that there can be long gaps in policy adoption between municipalities in the Netherlands (Jans et al., 2016). Future studies could replicate the study at a later moment to investigate the adoption of resilience policy over time. This research can be a starting point for future resilience policy research that utilizes time-series data.

As mentioned in this study, cities have interpreted the concept of resilience and its connected policies according to their own needs (Wang, 2021). Since this study did not go into detail about the resilience plans or the integration of the concept into policy by looking at the specific policies, no claims about effectiveness of the policy can be made. Future studies could look at the municipalities that have adopted a resilience plan and the policies that have been implemented as a result, in order to assess its effectiveness.

Nonetheless, even if structural stress is absent from a city at the current moment, the shocks and stresses resilience seeks to prepare for are unexpected and could affect any city. All cities therefore could benefit from preparing for such an event and adopting resilience into their policy or creating a resilience strategy. It is therefore important that multinational actors such as the United Nations, which have set up programs to assist cities in becoming resilient and adopt resilience policy, specifically target cities that do not have ambitions to become globalized and are not part of any networks yet. They should also target cities that do not currently face structural stress. Scholars have identified the lack of institutional incentives as a possible factor for municipalities lagging in adopting resilience policy (Coaffee et al., 2018; Huck et al., 2020; Lee, 2013). For that reason, institutional incentives or pressure from the national government could help more cities adopt policies aiming to make cities more resilient.

6 Conclusion

This research project employed mixed methods to provide insight into the internal and external drivers of policy adoption and their relative importance. The study focused on the 50 largest municipalities in the Netherlands as case studies in this process. The results suggest that internal drivers, specifically problem pressure, are more important for policy adoption when it comes to resilience policy than external drivers, specifically network participation. Certain factors that the existing academic literature commonly identifies as important to policy adoption, such as resources and network participation, have not been found significant in this study. Aligning with the literature, the problem pressure on cities is found to be significant and positively correlated with the likeliness of resilience policy adoption.

Despite the limited adoption of resilience as a concept by municipalities so far, the data suggest that the cities that could benefit most from resilience policy, those experiencing a high problem pressure, are in fact more likely to have adopted resilience policy. The possible implications of these findings are that resilience policy and the push by multilateral organizations have been effectively targeting cities that can especially benefit from adopting resilience policy as they experience structural stress and are more at risk. In contrast with the literature, according to the findings of this study, a city's level of globalization and resources do not play a role in the likelihood of integration of resilience into policy. A possible explanation is the small sample size of this study and the focus on the largest municipalities in the Netherlands.

As stated before, the risks cities face due to climate change, along with their contributions to it, demonstrate the benefits of the adoption of the concept of resilience to cities. As resilience policy research in general and policy diffusion research specifically focusing on resilience policy is limited, more research can help make resilience policy and the integration of the concept of resilience into policy more mainstream. This research project sought to contribute to making the adoption of resilience policy more common.

A little less than half (23) of the 50 municipalities have adopted an actual resilience plan, and only 3 cities have adopted an interdisciplinary plan. There are several reasons municipalities could have chosen not to integrate resilience into their policy. These are out of the scope of this study. One of the main difficulties municipalities face when implementing overarching multidisciplinary resilience policy is having to work across different departments (Coaffee et al., 2018). This points to the interdisciplinary characteristics of resilience policy. Multinational organizations or national-level institutions could help break down these barriers.

References

Abel, D. (2021). The diffusion of climate policies among German municipalities. *Journal of Public Policy, 41*(1), 111–136. https://doi.org/10.1017/s0143814x19000199.

Aguiar, F. C., Bentz, J., Silva, J. M. N., Fonseca, A. L., Swart, R., Santos, F. D., & Penha-Lopes, G. (2018). Adaptation to climate change at local level in Europe: An overview. *Environmental Science & Policy, 86*, 38–63. https://doi.org/10.1016/j.envsci.2018.04.010.

Araos, M., Ford, J., Berrang-Ford, L., Biesbroek, R., & Moser, S. (2016). Climate change adaptation planning for global south megacities: The case of Dhaka. *Journal of Environmental Policy & Planning, 19*(6), 682–696. https://doi.org/10.1080/1523908X.2016.1264873.

Bellinson, R., & Chu, E. (2019). Learning pathways and the governance of innovations in urban climate change resilience and adaptation. *Journal of Environmental Policy & Planning, 21* (1), 76–89. https://doi.org/10.1080/1523908X.2018.1493916.

Bergero, C., Rich, M., & Saikawa, E. (2021). All roads lead to Paris: The eight pathways to renewable energy target adoption. *Energy Research & Social Science, 80*. https://doi.org/10.1016/j.erss.2021.102215.

Berry, F. S., & Berry, W. D. (2007). Innovation and diffusion models in policy research. In P. A. Sabatier, & C. M. Weible (Eds.), *Theories of the policy process* (2nd ed., pp. 307–362). Westview Press.

Blatter, J., Portmann, L., & Rausis, F. (2021). Theorizing policy diffusion: From a patchy set of mechanisms to a paradigmatic typology. *Journal of European Public Policy, 29*(6), 805–825. https://doi.org/10.1080/13501763.2021.1892801.

Braverman, M. T., Geldhof, G. J., Hoogesteger, L. A., & Johnson, J. A. (2018). Predicting students' noncompliance with a smoke-free university campus policy. *Preventive Medicine, 114*, 209–216. https://doi.org/10.1016/j.ypmed.2018.07.002.

Calliari, E., Castellari, S., Davis, M. K., et al. (2022). Building climate resilience through nature-based solutions in Europe: A review of enabling knowledge, finance and governance frameworks. *Climate Risk Management, 37*, 100450.

Carley, S., & Nicholson-Crotty, S. (2018). Moving beyond theories of neighborly emulation: Energy policy information channels are plentiful among American states. *Energy Research & Social Science, 46*, 245–251. https://doi.org/10.1016/j.erss.2018.07.026.

CBS. (2022). *Bevolkingsontwikkeling; regio per maand [Data set]*. Statline https://opendata.cbs.nl/statline/#/CBS/nl/dataset/37230NED/table?fromstatweb.

Coaffee, J., Therrien, M. C., Chelleri, L., Henstra, D., Aldrich, D. P., Mitchell, C. L., Tsenkova, S., & Rigaud, É. (2018). Urban resilience implementation: A policy challenge and research agenda for the 21st century. *Journal of Contingencies & Crisis Management, 26*(3), 403–410. https://doi.org/10.1111/1468-5973.12233.

Danaeefard, H., & Mahdizadeh, F. (2022). Public policy diffusion: A scoping review. *Public Organization Review, 22*(2), 455–477. https://doi.org/10.1007/s11115-022-00618-9.

De Ruiter, R., & Schalk, J. (2017). Explaining cross-national policy diffusion in national parliaments: A longitudinal case study of plenary debates in the Dutch parliament. *Acta Politica, 52*(2), 133–155. https://doi.org/10.1057/ap.2015.29.

DellaVigna, S., & Kim, W. (2022). *Policy diffusion and polarization across U.S. states.* (National Bureau of Economic Research Working Paper 30142).

Dorst, H., Van der Jagt, A., Raven, R., & Runhaar, H. (2019). Urban greening through nature-based solutions – Key characteristics of an emerging concept. *Sustainable Cities and Society, 49*, 101620.

Dorst, H., van der Jagt, A., Toxopeus, H., Tozer, L., Raven, R., & Runhaar, H. (2022). What's behind the barriers? Uncovering structural conditions working against urban nature-based solutions. *Landscape and Urban Planning, 220*, 104335.

Eraydin, A., & Özatağan, G. (2021). Pathways to a resilient future: A review of policy agendas and governance practices in shrinking cities. *Cities, 115*. https://doi.org/10.1016/j.cities.2021.103226.

FinDo. (2022). *Data Financiën Decentrale Overheden [Data set]*. FinDo https://findo.nl/dashboard/dashboard/gemeentelijke-begroting.

Fuentes, A., & Pipkin, S. (2022). Appetite for reform: When do exogenous shocks motivate industrial policy change? *The Journal of Development Studies, 58*(6), 1081–1101. https://doi.org/10.1080/00220388.2021.2017890.

Fünfgeld, H. (2014). Facilitating local climate change adaptation through transnational municipal networks. *Current Opinion in Environmental Sustainability, 12*. https://doi.org/10.1016/j.cosust.2014.10.011.

G40. (n.d.). *41 Steden. G40*. https://www.g40stedennetwerk.nl/41-steden.

Gilardi, F., & Wasserfallen, F. (2019). The politics of policy diffusion. *European Journal of Political Research*, *58*(4), 1245–1256. https://doi.org/10.1111/1475-6765.12326.

Gordon, D. J. (2013). Between local innovation and global impact: Cities, networks, and the governance of climate change. *Canadian Foreign Policy Journal*, *19*(3), 288–307.

Haupt, W., Chelleri, L., Van Herk, S., & Zevenbergen, C. (2020). City-to-city learning within climate city networks: Definition, significance, and challenges from a global perspective. *International Journal of Urban Sustainable Development*, *12*(2), 143–159. https://doi.org/10.1080/19463138.2019.1691007.

Haupt, W., & Coppola, A. (2019). Climate governance in transnational municipal networks: Advancing a potential agenda for analysis and typology. *International Journal of Urban Sustainable Development*, *11*(2), 123–140. https://doi.org/10.1080/19463138.2019.1583235.

Häußler, S., & Haupt, W. (2021). Climate change adaptation networks for small and medium-sized cities. *SN Social Sciences*, *1*. https://doi.org/10.1007/s43545-021-00267-7.

Heikkinen, M. (2022). The role of network participation in climate change mitigation: A city-level analysis. *International Journal of Urban Sustainable Development*, *14*(1), 1–14. https://doi.org/10.1080/19463138.2022.2036163.

Heikkinen, M., Aasa, K., Klein, J., Juhola, S., & Ylä-Anttila, T. (2020). Transnational municipal networks and climate change adaptation: A study of 377 cities. *Journal of Cleaner Production*, *257*. https://doi.org/10.1016/j.jclepro.2020.120474.

Henley, T., & Lee, Y. (2023). Pressures to local governments for unrequired ethics policy adoption. *Public Administration Quarterly*, *47*(1), 1–25. https://doi.org/10.37808/paq.47.1.1.

Huck, A., Monstadt, J., & Driessen, P. (2020). Mainstreaming resilience in urban policy making? insights from Christchurch and Rotterdam. *Geoforum*, *117*, 194–205. https://doi.org/10.1016/j.geoforum.2020.10.001.

Huck, A., Monstadt, J., Driessen, P. P., & Rudolph-Cleff, A. (2021). Towards resilient Rotterdam? Key conditions for a networked approach to managing urban infrastructure risks. *Journal of Contingencies & Crisis Management*, *29*(1), 12–22. https://doi.org/10.1111/1468-5973.12295.

Jänicke, M. (2005). Trend-setters in environmental policy: The character and role of pioneer countries. *Environmental Policy and Governance*, *15*(2), 129–142. https://doi.org/10.1002/eet.375.

Jans, W., Denters, B., Need, A., & Van Gerven, M. (2016). Mandatory innovation in a decentralised system: The adoption of an e-government innovation in Dutch municipalities. *Acta Politica*, *51*(1), 36–60. https://doi.org/10.1057/ap.2014.36.

Kabisch, N., Frantzeskaki, N., & Hansen, R. (2022). Principles for urban nature-based solutions. *Ambio*, *51*(6), 1388–1401.

Kammerer, M., & Namhata, C. (2018). What drives the adoption of climate change mitigation policy? A dynamic network approach to policy diffusion. *Policy Sciences*, *51*(4), 477–513. https://doi.org/10.1007/s11077-018-9332-6.

Kincaid, D. L. (2004). From innovation to social norm: Bounded normative influence. *Journal of Health Communication*, *Suppl 1*, 37–57. https://doi.org/10.1080/10810730490271511.

Kousky, C., & Schneider, S. H. (2003). Global climate policy: Will cities lead the way? *Climate Policy*, *3*(4), 359–372. https://doi.org/10.1016/j.clipol.2003.08.002.

Kuhlmann, J., González de Reufels, D., Schlichte, K., & Nullmeier, F. (2019). How social policy travels: A refined model of diffusion. *Global Social Policy*, *20*(1), 80–96. https://doi.org/10.1177/1468018119888443.

Lee, T. (2013). Global cities and transnational climate change networks. *Global Environmental Politics*, *13*(1), 108–127. https://doi.org/10.1162/GLEP_a_00156.

Lee, T., & Koski, C. (2014). Mitigating global warming in global cities: Comparing participation and climate change policies of C40 cities. *Journal of Comparative Policy Analysis: Research and Practice, 16*(5), 475–492. https://doi.org/10.1080/13876988.2014.910938.

Lee, T., & van de Meene, S. (2012). Who teaches and who learns? Policy learning through the C40 cities climate network. *Policy Sciences, 45,* 199–220. https://doi.org/10.1007/s11077-012-9159-5.

Leidelmeijer & Mandemakers. (2022). *Leefbaarheid in Nederland 2020.* Atlas Research. https://open.overheid.nl/documenten/ronl-dad15f458d26412cb342bc966c870ac7f1ff77fd/pdf.

Lu, P., & Stead, D. (2013). Understanding the notion of resilience in spatial planning: A case study of Rotterdam, The Netherlands. *Cities, 35,* 200–212. https://doi.org/10.1016/j.cities.2013.06.001.

M50. (n.d.). *Over M50.* M50. https://www.middelgrotegemeenten.nl/strag+m50/default.aspx.

Mallinson, D. J. (2021). Policy innovation adoption across the diffusion life course. *Policy Studies Journal, 49*(2), 335–358. https://doi.org/10.1111/psj.12406.

Massey, E., Biesbroek, R., Huitema, D., & Jordan, A. (2014). Climate policy innovation: The adoption and diffusion of adaptation policies across Europe. *Global Environmental Change, 29,* 434–443. https://doi.org/10.1016/j.gloenvcha.2014.09.002.

Massey, E., & Huitema, D. (2013). The emergence of climate change adaptation as a policy field: The case of England. *Regional Environmental Change, 13,* 341–352. https://doi.org/10.1007/s10113-012-0341-2.

Matisoff, D. C. (2008). The adoption of state climate change policies and renewable portfolio standards: Regional diffusion or internal determinants? *Review of Policy Research, 25,* 527–546. https://doi.org/10.1111/j.1541-1338.2008.00360.x.

Matisoff, D. C., & Edwards, J. (2014). Kindred spirits or intergovernmental competition? The innovation and diffusion of energy policies in the American states (1990–2008). *Environmental Politics, 23*(5), 795–817. https://doi.org/10.1080/09644016.2014.923639.

Mohr, L. B. (1969). Determinants of innovation in organizations. *The American Political Science Review, 63*(1), 111–126. https://doi.org/10.2307/1954288.

MRA. (n.d.). *Over de Metropoolregio Amsterdam.* Metropoolregio Amsterdam. https://www.metropoolregioamsterdam.nl/over-mra/.

MRDH. (n.d.). *Ambities.* Metropoolregio Rotterdam Den Haag. https://mrdh.nl/wie-zijn/ambities.

MRE. (n.d.). *Samenwerken is samenwerken.* Metropoolregio Eindhoven. *https://metropoolregioeindhoven.nl/over-ons/samenwerken-is-samen-werken.*

Nguyen, T. M. P., Davidson, K., & Coenen, L. (2020). Understanding how city networks are leveraging climate action: Experimentation through C40. *Urban Transformations, 2*(12). https://doi.org/10.1186/s42854-020-00017-7.

O'Donnell, E. C., Lamond, J. E., & Thorne, C. R. (2018). Learning and action alliance framework to facilitate stakeholder collaboration and social learning in urban flood risk management. *Environmental Science & Policy, 80,* 1–8.

Obinger, H., Schmitt, C., & Starke, P. (2013). Policy diffusion and policy transfer in comparative welfare state research. *Social Policy & Administration, 47*(1), 111–129. https://doi.org/10.1111/spol.12003.

OECD. (n.d.-a). *Cities and environment.* OECD. https://www.oecd.org/regional/cities/cities-environment.htm.

OECD. (n.d.-b). *Resilient cities.* OECD. https://www.oecd.org/cfe/resilient-cities.htm.

Orellana, S. (2010). How electoral systems can influence policy innovation. *Policy Studies Journal, 38,* 613–628. https://doi.org/10.1111/j.1541-0072.2010.00376.x.

Otto, A., Kern, K., Haupt, W., Eckersley, P., & Thieken, A. H. (2021). Ranking local climate policy: Assessing the mitigation and adaptation activities of 104 German cities. *Climatic Change*, *167*, 1–2. https://doi.org/10.1007/s10584-021-03142-9.

Papin, M. (2019). Transnational municipal networks: Harbingers of innovation for global adaptation governance? *International Environmental Agreements: Politics, Law and Economics*, *18*, 467–483. https://doi.org/10.1007/s10784-019-09446-7.

Pereira, M. M. (2022). How do public officials learn about policy? A field experiment on policy diffusion. *British Journal of Political Science*, *52*(3), 1428–1435. https://doi.org/10.1017/s0007123420000770.

Rashidi, K., & Patt, A. (2018). Subsistence over symbolism: the role of transnational municipal networks on cities' climate policy innovation and adoption. *Mitigation and Adaptation Strategies for Global Change*, *23*, 507–523. https://doi.org/10.1007/s11027-017-9747-y.

Rogers, E. M. (1983). *Diffusion of innovations*. Free Press.

Ryan, D. (2015). From commitment to action: A literature review on climate policy implementation at city level. *Climatic Change*, *131*, 519–529. https://doi.org/10.1007/s10584-015-1402-6.

Saikawa, E. (2013). Policy diffusion of emission standards is there a race to the top? *World Politics*, *65*(1), 1–33. https://doi.org/10.1017/S0043887112000238.

Salvia, M., Reckien, D., Pietrapertosa, F., Eckersley, P., Spyridaki, N., Krook-Riekkola, A., Olazabal, M., De Gregorio Hurtado, S., Simoes, S. G., Geneletti, D., Viguié, V., Fokaides, P. A., Ioannou, B. I., Flamos, A., Csete, M. S., Buzasi, A., Orru, H., de Boer, C., Foley, A., ... Heidrich, O. (2021). Will climate mitigation ambitions lead to carbon neutrality? An analysis of the local-level plans of 327 cities in the EU. *Renewable and Sustainable Energy Reviews*, *135*. https://doi.org/10.1016/j.rser.2020.110253.

Sarabi, E. S., Han, Q., Romme, A. G. L., de Vries, B., & Wendling, L. (2019). Key enablers of and barriers to the uptake and implementation of nature-based solutions in urban settings: A review. *Resources*, *8*, 121. https://doi.org/10.3390/resources8030121.

Schoenefeld, J. J., Schulze, K., & Bruch, N. (2022). The diffusion of climate change adaptation policy. *WIREs Climate Change*, *13*(3). https://doi.org/10.1002/wcc.775.

Shamsuddin, S. (2020). Resilience resistance: The challenges and implications of urban resilience implementation. *Cities*, *103*. https://doi.org/10.1016/j.cities.2020.102763.

Sharma, S. E. (2022). Urban flood resilience: Governing conflicting urbanism and climate action in Amsterdam. *Review of International Political Economy*, *1–23*. https://doi.org/10.1080/09692290.2022.2100449.

Sharma, M., Sharma, B., Kumar, N., & Kumar, A. (2023). Establishing conceptual components for urban resilience: Taking clues from urbanization through a planner's lens. *Natural Hazards Review*, *24*(1). https://doi.org/10.1061/NHREFO.NHENG-1523.

Shipan, C. R., & Volden, C. (2008). The mechanisms of policy diffusion. *American Journal of Political Science*, *52*(4), 840–857. https://doi.org/10.1111/j.1540-5907.2008.00346.x.

Short, J. R. (2015 July 31). *How cities are adapting to climate change*. World Economic Forum. https://www.weforum.org/agenda/2015/07/how-cities-are-adapting-to-climate-change/.

Stewart, G., Kamata, A., Miles, R., Grandoit, E., Mandelbaum, F., Quinn, C., & Rabin, L. (2019). Predicting mental health help seeking orientations among diverse undergraduates: An ordinal logistic regression analysis. *Journal of Affective Disorders*, *257*. https://doi.org/10.1016/j.jad.2019.07.058.

Urban Resilience Hub. (n.d.). *Building resilience*. Urban Resilience Hub. https://urbanresiliencehub.org/building-resilience/.

Verschueren, C. (2022). Multi-layered predictors of ESE policy adoption in large school districts in the United States. *Environmental Education Research*, *28*(8), 1251–1270. https://doi.org/10.1080/13504622.2022.2080184.

VNG. (n.d.-a). *Netwerken van onze leden*. Vereniging van Nederlandse Gemeenten. https://vng.nl/artikelen/netwerken-van-onze-leden.

VNG. (n.d.-b) *Overzichtskaart van gemeenten die meedoen aan*. Global Goals. Vereniging van Nederlandse Gemeenten. https://vng.nl/artikelen/overzichtskaart-van-gemeenten-die-meedoen-aan-global-goals.

Walker, J. (1969). The diffusion of innovations among the American States. *American Political Science Review*, *63*(3), 880–899. https://doi.org/10.2307/1954434.

Wang, C. (2021). Research on sponge city planning based on resilient city concept. *IOP Conference Series: Earth and Environmental Science*, *793*(1). https://doi.org/10.1088/1755-1315/793/1/012011.

Yi, H., & Liu, I. (2022). Executive leadership, policy tourism, and policy diffusion among local governments. *Public Administration Review*, *82*(6), 1024–1041. https://doi.org/10.1111/puar.13529.

Zebrowski, C. (2020). Acting local, thinking global: Globalizing resilience through 100 resilient cities: Review of Central European affairs. *New Perspectives*, *28*(1), 71–88. https://doi.org/10.1177/2336825X20906315.

Zhang, Z., & Zhu, X. (2019). Multiple mechanisms of policy diffusion in China. *Public Management Review*, *21*(4), 495–514. https://doi.org/10.1080/14719037.2018.1497695.

Zhou, S., Matisoff, D., Kingsley, G., & Brown, M. (2019). Understanding renewable energy policy adoption and evolution in Europe: The impact of coercion, normative emulation, competition, and learning. *Energy Research & Social Science*, *51*. https://doi.org/10.1016/j.erss.2018.12.011.

Zingraff-Hamed, A., Hüesker, F., Albert, C., Brillinger, M., Huang, J., Lupp, G., Scheuer, S., Schlätel, M., & Schröter, B. (2021). Governance models for nature-based solutions: Seventeen cases from Germany. *Ambio*, *50*(8), 1610–1627. https://doi.org/10.1007/s13280-020-01412-x.

Index

Note: Page numbers followed by *f* indicate figures and *t* indicate tables.

Printed in the United States
by Baker & Taylor Publisher Services